## ABOUT THE

INGRID BETANCOURT, a senator and a presidential candidate for Colombia at the time of her kidnapping, grew up among diplomats, literati, and artists who congregated at her parents' elegant home in Paris. Her father served as Colombia's ambassador to UNESCO, and her mother, a political activist, continued her work on behalf of the country's countless children whose lives were being destroyed by extreme poverty and institutional neglect. Intellectually, Ingrid was influenced by Pablo Neruda and other Latin American writers such as Gabriel García Márquez, who frequented her parents' social circle. She studied at École de Sciences Politiques de Paris, a prestigious academy in France.

From this charmed life, Ingrid Betancourt—not yet thirty, happily married to a French diplomat, and a mother to two children—returned to her native country in the late eighties. On what was initially just a visit, she found her country under internal siege from the drug cartels and the corrupt government that had allowed them to flourish. After seeing what had become of Colombia's democracy, she could not leave.

*Until Death Do Us Part* is the deeply personal story of a woman who gave up a life of comfort and safety to become a political leader in a country being slowly demolished by terrorism, violence, fear, and a pervasive sense of hopelessness. It is a country where democracy is being sacrificed for the well-being of the few, where international criminals determine policy, and where political assassinations are a way of life. Ingrid Betancourt was elected and reelected as a representative and as a senator in Colombia's national legislature. She founded a political party that has openly confronted Colombia's leaders and earned the respect of a nation. And then she became a target.

Forced to move her children out of Colombia for protection against death threats, Ingrid Betancourt remained and continued to fight the political structure that had crumbled under the destructive power of the paramilitary forces, the financial omnipotence of the drug cartels, and the passivity of government-for-sale until her kidnapping by the Revolutionary Armed Forces of Colombia (FARC) in 2002.

*Until Death Do Us Part* chronicles a life spent fighting the political establishment. It is not only a chilling account of the dangerous, byzantine machine that runs Colombia but also an inspiring story of privilege, sacrifice, and true patriotism.

On July 2, 2008, Ingrid Betancourt was rescued by the Colombian military.

# UNTIL DEATH DO US PART

AN **ECCO** BOOK

HARPER  **PERENNIAL**

NEW YORK • LONDON • TORONTO • SYDNEY • NEW DELHI • AUCKLAND

# UNTIL DEATH DO US PART

## MY STRUGGLE TO RECLAIM COLOMBIA

### INGRID
### BETANCOURT

*Translated from the French by Steven Rendall*

HARPER ● PERENNIAL

A hardcover edition of this book was published in 2002 by Ecco, an imprint of HarperCollins Publishers.

FIRST HARPER PERENNIAL EDITION PUBLISHED 2008.

*Map designed by Jackie Aher*
*Designed by Gretchen Achilles*

The Library of Congress has catalogued the hardcover edition as follows:

Betancourt, Ingrid, 1961–
    [Rage au cœur. English]
    Until death do us part : my struggle to reclaim Colombia / Ingrid Betancourt ; translated from the French by Steven Rendall.—1st ed.
    p. cm.
    ISBN 0-06-000890-3 (hardcover)
    1. Betancourt, Ingrid, 1961– 2. Colombia—History—1974– 3. Political corruption—Colombia—History. 4. Drug traffic—Colombia—History. 5. Presidential candidates—Colombia—Biography. 6. Legislators—Colombia—Biography. I. Title.

F2279.22.B48 A3 2002
986.106'35'092—dc21
[B]                                                                                   2001040691

ISBN 978-0-06-000891-8 (pbk.)

08 09 10 11 12  RRD  10 9 8 7 6 5 4 3 2 1

TO MELANIE AND LORENZO,

WHO HAVE MADE

IT ALL WORTHWHILE

# PUBLISHER'S NOTE

Ingrid Betancourt was taken hostage by the FARC on February 23, 2002. At the time, she was campaigning to become president of Colombia on an anticorruption ticket in one of the most corrupt countries in the world. As an elected representative and then senator of the Colombian legislature, her love for her country and her faith in democracy gave her the courage to stand up to the power that has subjugated, intimidated, and corrupted all those who oppose it. For six and a half years, she was held captive in the depths of the Colombian jungles.

On July 2, 2008, she was rescued along with fourteen other hostages by Colombian soldiers who were posing as members of a nongovernment organization. Her rescue made news around the globe.

*Until Death Do Us Part* was originally published in America weeks before Ingrid Betancourt was kidnapped by the FARC. Her message of hope, courage, and change is beautifully captured in these pages.

# UNTIL DEATH DO US PART

# PROLOGUE

**DECEMBER 1996.** Vacation begins in a few days; the legislative session is almost over. Even more than usual, I'm rushing back and forth between my office, where I have back-to-back appointments, and the legislative assembly, where I'm supposed to speak. I'm thirty-five years old, and I've been a member of the legislature for two years.

Toward three-thirty in the afternoon, while I'm talking with someone in my office, my secretary pokes her head in the door.

"Someone's asking to see you right away, Ingrid. A man."

"Does he have an appointment?"

"No. But he's very insistent."

The debate in the assembly starts at four. I think for a moment.

"All right, tell him I'll see him immediately after this person, but for no more than half an hour. That's all the time I have."

He comes in: elegant, in his forties, average height, neither handsome nor ugly, so that later on I will be unable to describe or identify him.

"Please sit down."

"Thank you. We've been following your work with the greatest attention, *Doctora*, and we have the highest regard for what you're doing."

We smile at each other. I sit erect, with my elbows on the desk that separates us; I assume he's going to ask for something, like most of the people who come to see me.

"And that's why I wanted to meet you, *Doctora*. We're very worried about you. Colombia is going through a period of great tension, great violence. One must be careful, very, very careful."

Then he frowns, grows more serious, stops looking me in the eyes.

I'm used to this kind of talk. Most of the people I meet and who support me share this obsession with danger. Women, in particular, invariably assure me, with genuine affection, that they're praying that nothing happens to me, that God will protect me. I try to convince them that my security is very tight and I'm in no danger, because I believe those in power exploit this fear that grips Colombians. What better way to destroy people's hopes than to persuade them that anyone who dares to speak, to accuse, will inevitably be eliminated?

"Don't worry," I tell this man, "I'm very well protected. I'm surrounded by a discreet but highly effective security apparatus; there's nothing to fear. That said, I'm grateful for your interest in my welfare. But what can I do for you?"

Surprisingly, he repeats what I'd taken to be a polite introduction to a request, his eyes a little steelier.

"I'd like to know you better, *Doctora*, but the reason I'm here is to warn you. We are extremely concerned."

"That's very kind of you, and I'm touched by your concern, but I have very little time, as my secretary must have told you." I look at my watch, making sure he sees me do it.

"You haven't understood me," he continues coldly. "I'm telling you that you must really be careful."

This time there's nothing friendly in his face. He sits there and looks at me fixedly. I realize that he's not the kind of visitor I'd imagined, not a citizen in distress who's come to ask for help, or a bashful admirer, but an emissary with a very specific message for me. I also change my tone.

"What's the message?" I ask, with a slight laugh. "You want to give me a message: What is it? Are you threatening me?"

"No, this is not a threat. I'm not here to frighten you. You have to realize that you're in danger, that your family is in danger. I'm speaking to you on behalf of people who've already put out a contract on you. They advise you to leave, because the decision has already been made. To be perfectly clear, what I'm telling you, *Doctora*, is that we've already paid the *sicarios*."

I feel the blood drain from my face. Suddenly, I know he's not lying. In Colombia, the word *sicario* makes everything clear. *Sicarios* are young men with motorcycles who live in Colombia's poorest neighborhoods, and they're hired every day to kill people for ridiculously small sums of money.

I've turned a corner, crossed a red line, and this time the period of intimidation is really over. Six months earlier, as I was leaving the Capitol on a cold night in July, shots were fired at my car and that of my bodyguards. No one was hit, and I tried to believe that we'd just been at the wrong place at the wrong time.

"In short, what you're telling me," I say, articulating my words clearly and looking him straight in the eye, "is that you're going to kill me."

"I've come to tell you to leave because steps have already been taken."

He gets up, holds out his hand, politely says goodbye, and leaves. Did I shake his hand? Did I even smile back at him? It's entirely possible that I did. I no longer remember.

Now I'm alone in my office, stunned, drained, inert. A few seconds pass before I recover my wits and the strength to call my secretary.

"Marina, where did that guy come from? How did he get in?"

"I don't know. All of a sudden he was just there, in my office."

"What's his name? Did you get his name, at least?"

"No. I thought he knew you, that he was one of your friends."

One doesn't get into the Capitol without being invited by a member, presenting one's papers, or giving one's name. But he just sailed right in and entered my office without anyone asking him a single question.

Whom should I notify? The police? They're on the payroll of the same government that wants to shut me up—in particular, the head of state, Ernesto Samper. For months, I've been almost the only one denouncing his corruption. On the other hand, my visitor may well be part of the security service, which would explain how he was able to make his way through the building without being stopped.

I sink for a few instants into visions of darkest nightmares. I have no one to protect me. They might kill us soon, maybe this evening. He said: "Your family is in danger." Melanie and Lorenzo, my children; Juan Carlos, with whom I live. Who can I call upon for help? I have no one, no recourse, no way of saving them, no way of eliminating the threat that hangs over them. Somewhere in Bogotá men have been paid and armed; they might attack us at any time.

Pick up the children, right away! Melanie is eleven, Lorenzo only seven. Loli, my baby . . . They go to the French school; that's no secret, anyone could find it out by asking my concierge or our neighbors. Anyone at all. My driver drops them there every morning and picks them up in the evening—or I do, when I can. I'm constantly accompanied by my bodyguards, but they themselves have no protection. Yes, go get the kids, immediately. Every hour, every minute that goes by is heavy with an unutterable, unimaginable anxiety . . .

"Marina, I have to leave, it's urgent. Do the best you can, I'll call you tomorrow."

I'm breathless; I leave everyone in the lurch. Running through the endless corridors of the Capitol, crossing the checkpoints, the heavy doors. If only . . . Yes, my driver's there, discreetly parked at the corner of the Bolivar Plaza.

He sees me, starts toward me. I trust him completely; we've lived through moments of fear together, and it was perhaps because of him, because of his skill, his presence of mind, that we escaped the shots aimed at us six months earlier.

"The children, Alex. Hurry! Hurry! I'll explain later."

Poor Alex! It's rush hour, the offices are beginning to close, and six million residents of Bogotá are storming the terrible buses that lack doors or windows and spew choking black smoke. At the end of the millennium, they are all our disgraceful political class has left us in the way of public transportation. Bogotá has neither subways nor trams, only broad, potholed avenues that are jammed at this time of afternoon. Alex knows how to deal with this situation; he forces his way through, ignoring the honking horns and the insults, and my escort follows.

Now I have to warn Juan Carlos and tell him to join us. Juan Carlos is only slightly older than I, but he's solid, serene. At the worst times over the past year, he's always been at my side to advise me, to comfort me, and sometimes to protect me.

"Juan Carlos! It's me. It's very serious. We've got to talk right away. Can you come?"

"Where are you?"

"In the car. I'm going to pick up the kids and then go home."

"I'll be there in half an hour. Be careful."

The traffic is moving more smoothly now. The French school is near the French embassy, in the center of the northern part of the city, the sanctuary of Bogotá's upper class. Here, the walls surrounding the luxurious houses are topped by rows of surveillance cameras, and there are even armed guards wearing bulletproof vests.

Loli, at last! Loli, called out of his classroom, surprised, his hair mussed, his book bag only half buckled, books and papers poking out of it.

"Loli!"

"Are you all right, Mama?"

"Yes, I'm all right. I just wanted all of us to spend the evening together for a change. I was able to get away."

And then Melanie, who looks like me, but more luminous and well groomed.

"Hey, what are you doing here, Mama? I thought we were going to see you tonight."

"I changed my mind. We're going to make plans for the vacation. Kiss me, Mela. Loli, give me your book bag."

He is talking to me about an activity or show his class is preparing for Christmas, but I've already stopped listening. I watch Alex open the doors and tenderly tell the children to get in. My eyes instinctively sweep the street. Lord, don't let a motorcycle come along! I don't care about cars; the *sicarios* don't drive around in cars.

"Be especially careful of guys on motorcycles, Alex, okay? Now take us home as quickly as you can."

He laughs.

"What guys on motorcycles? Don Juan Carlos rides a motorcycle!"

"Yes, that's right, I'm an idiot. Excuse me."

Juan Carlos travels only by motorcycle, I remind myself, so not all motorcyclists are killers.

# CHAPTER ONE

**THE CHILDREN** are eating their snack in the kitchen. We hear them laughing. In the adjoining room, I tell Juan Carlos what the man said. His words are precisely engraved on my memory, right down to their rhythm and intonation. In the meantime, they've taken on a terrifying, permanent, unforgettable meaning.

"We have to get the children out of here, Ingrid. Immediately."

"Yes."

"Call their father in New Zealand and tell him we're bringing them on the first plane we can get."

Juan Carlos says out loud what I already know, what I decided during the interminable trip from the Capitol to the French school. He can't imagine how much it helps me to hear him express what is for me the worst possible horror: they have to leave. They have to leave for a long time, I know it—for years, maybe. To save their lives, I have to make them leave. Juan Carlos says what has to be done, but with his eyes he silently tells me he'll be there to help me bear this unheard-of burden: their absence, the void, the gulf on the edge of which we'll have to live from now on. That he'll be there.

Not for a second does he suggest that I abandon my battle against government corruption. For the time being, this amounts to hardly more than a handful of sand thrown into the monstrous gears of a machine that has ground up the few heedless people who have challenged it. I think of my mother's close friend, Luis Carlos Galán, who was a candidate for the presidency of Colombia and was assassinated at the beginning of an electoral meeting in 1989. He was forty-six years old, and my mother was at his bedside when he died. I wanted to take up the torch, and Colombians heard me when they elected me to the legislature in 1994 with more votes than any other candidate in the Liberal Party, Galán's party. I'll go all the way for the Colombian people, whom our political class has despised and robbed, generation after generation. I won't give up, whatever price has to be paid. This evening, I'm grateful to Juan Carlos for not doubting my resolve, not challenging this commitment.

Fabrice, the father of my children, is French, a diplomat currently posted in Auckland. We separated in 1990, and Colombia played a large part in our separation. But once the effects of the split had dissipated, a strong, special friendship formed between us, and we have recovered the esteem we had for each other.

"Did something happen? Were they threatened?"

"Threatened, yes. Nothing more. They're fine, they're right here, don't worry, but I can't wait—they have to leave."

"For good, you mean?"

"For a long time. I can't explain everything here on the phone. I need your help."

"All right. Come on the first plane you can get . . . Ingrid? Are you going to be all right? You're not all alone?"

"Juan Carlos is here, he'll be traveling with us."

Now I have to speak to the children, while Juan Carlos finds us seats on an international flight. It doesn't matter where it's going, just as long as we get out of Colombia. We'll find a way to get to Auckland later.

"Melanie, Loli, listen to me, I've got something important to tell you. We're going to spend Christmas in Auckland."

"With Papa?"

"Yes, that's right."

"Great!"

"Yes, my darling, it's great. But we've got to leave sooner than planned."

"Before school is out?"

"Tomorrow morning, in fact."

"We can't do that! We left all our stuff."

"We'll notify the school, Melanie, don't worry."

"So we're leaving just like that, without saying goodbye or anything? Why?"

"That's how it is, dear, I can't explain everything. We'll talk about it later if you want to, okay? Accept the situation as it is. It's a little rushed, I know, but it's good nonetheless, isn't it?"

"Yes, but . . ."

"And as far as your show's concerned, Loli, don't worry, I'll call. Okay, now let's get our bags packed."

It's done. We have four seats for Los Angeles tomorrow morning.

That night, Juan Carlos and I hardly sleep. We leave the light on, we listen for any unusual sound. For the first time, I have an imminent fear for my life, and that of my family's, as my visitor's message resounds in my mind. Over the past year, while the case against the president of Colombia, Ernesto Samper, was being prepared, I was feeling very much alone, fighting to bring it to its conclusion, to make public the proof of his guilt. Between August 1995 and March 1996, four witnesses for the prosecution were killed, one after another.* I kept the

---

*Dario Reyes, Horacio Serpa's driver, was killed on August 16, 1995. Alvaro Gómez Hurtado, leader from the Conservative Party, was killed on November 2, 1995. Elisabeth Montoya was killed on February 1, 1996. José Santacruz Londoño, the chief of the Cali Cartel, was killed on March 5, 1996.

newspapers with the police photos of those dark, closed faces. I'd met some of these witnesses, and I'm still haunted by their deaths. I want to bear testimony for them, too.

Though I'm usually confident in my strength, I feel fragile during these long hours, incredibly vulnerable, because this time I'm not the only one in the line of fire. The dreadful shadow that hangs over my children saps my resources, eats away at my heart.

I'm angry with myself for having chosen this apartment house at the foot of the mountain, at the end of a cul-de-sac. It's an ideal spot for an ambush: there's no way out. Not long ago a girl was kidnapped here, apparently without the slightest difficulty. To make things worse, my apartment is on the top floor, and thus accessible from the roof.

Auckland is a paradise compared with the black chaos of Bogotá. For a long time Auckland was a British possession, and the city is full of cottages surrounded by lawns. For us Colombians, who are constantly being pushed around and bowled over by the silent war that has been waged in our capital for decades, it's impossible to believe that a place like this exists, though we know it does.

It's high summer in the Southern Hemisphere. Fabrice, tanned and in shirtsleeves, is waiting for us at the airport. His face lights up, he holds out his arms, and the children run toward him. Just twenty-four hours ago, when we left home in the back of an armored car, terrified by the twenty minutes' drive to the airport, New Zealand was only a distant dream. Now Juan Carlos and I hold back to give them plenty of time. It's over, the children are no longer in danger, they're saved. We're numb with fatigue and emotion.

Fabrice has arranged things as best he could; he has moved in with some friends to leave us his house, to let us recover, reenter normal life. The house looks out on a garden full of flowers. It's spacious, blissfully calm. We spend our first moments there walking from room to room, wanting to laugh and cry at once, incredulous, unable to make any decisions at all. Then we give in to fatigue and sleep.

I haven't told my parents about our escape; I don't want to frighten them. For the past twenty years, they've both been living in Bogotá, but separately.

I call my mother. I hear myself explaining to her that I'm going to have to live without my children. After a few seconds she says:

"You know what, Ingrid? I'm coming to spend Christmas with you."

"Would you do that?"

"Of course. It'll be wonderful, you'll see."

We were supposed to celebrate Christmas together in Bogotá—well, too bad for Bogotá, the celebration will take place anyway. My mother, intelligent and generous, as she's always been, understands without further explanation.

As soon as I hang up, I call my father.

"It's settled, my dear. Stay were you are. We'll spend Christmas together; get a room ready for me, and I'll make a reservation."

Neither of them makes a single comment regarding my political commitment and the price, suddenly exorbitant, that I'll have to pay to stick to it. I know they share my suffering, but they tacitly support me. How could they prove that more fully than by making this long journey?

In Auckland, the days go by. We lead a family life that had become entirely foreign to us: picnics on the lawn, afternoons at the beach, evenings under the stars in the warm breeze off the Pacific. At night we go to bed without closing the doors or windows. The absence of the keys, fences, surveillance cameras, and bodyguards that constantly accompany me in Colombia contributes to the feeling that everything is unreal. This is not my life. It's a parenthesis, a precious reprieve of five or six weeks. I know that very well, so well that after a few days I can no longer fall asleep before six in the morning. The fear is there, lurking under my apparent lack of concern. Unable to relax, I sit up in bed and listen to the silence.

One night Juan Carlos finds me sitting up, and we start to talk. After that, until the end of our stay in Auckland, he remains at my side and we tell each other everything—our hopes, our dreams, and our fears—never drifting off to sleep before the first light of dawn.

I use these few magical weeks to construct an accelerated plan for my children's future. I give it all the attention of a mother who knows that she's not going to be there for many months. I meet all the teachers, buy books, notebooks, uniforms. Together we arrange their rooms, shop for clothes. Then I imprint places on my memory so I can imagine Melanie and Loli coming and going in this large, well-to-do town where no children sleep on the street, where the police force is there to protect citizens, where there are no *sicarios*. Their school—and this is an image I want to take away with me—is a charming green house in the middle of a garden. It seems that nothing bad can happen to schoolchildren who spend their time in this Eden.

We've said goodbye. Hugging my children at the airport, I suddenly see myself as my own mother, embracing my sister and me one last time before flying off to another continent. At a certain time in her life she also had to go away and leave us in the care of our father. What Melanie and Lorenzo are going to experience now—the discovery of another world and another language, the suffering of distance, departures, returns—my sister and I experienced many years ago. It played a significant role in our initiation into the world.

# CHAPTER TWO

MY FIRST MEMORIES go back to Neuilly, France. My father has rented a house on the edge of the Bois de Boulogne. I'm looking for lady-bugs in the garden; it's the early 1960s, and I'm two or three years old. In nursery school I speak French, but at home I hear languages that are spoken all over the world, depending on who my parents' guests are. My father is the assistant director of UNESCO, the United Nations Educational, Scientific and Cultural Organization, so there are many international guests at the house.

My elder sister, Astrid, and I are the beloved, pampered children of a refined couple who frequent the Parisian cultural world and whom most foreign artists visit when they're in town. My father is well past forty; he has already been minister of national education in Colombia, and in the chancelleries it's rumored that he might one day be president of Colombia. My mother is only twenty-five. She won a number of beauty contests in Bogotá before I was born, but she's better known in Colombia for her work with street children. Using the minor celebrity her beauty had won her, she forced her way into the ministry of justice and got the government to make available to her an unused prison in Bogotá, where she began housing children who'd been sleeping under bridges.

A shared passion for the welfare of children and young people played an important part in my parents' coming together. Gabriel Betancourt was a dedicated minister of education and still a bachelor when he met Yolanda Pulecio, whom people talked about because she'd opened the first *Albergue* (shelter) in the old prison and was now seeking other places she could convert into housing for children. My father also had a story to tell. In 1942, as a student, he had to ask a Colombian corporation for financial help to accomplish his dream of graduating from an American university. Once in the United States, he realized that many of his friends hadn't had the same opportunity. This is why he initiated a project for creating a student loan system, which, by the way, earned him his master's degree in public administration. Later, as minister of education, he developed his project and this program inspired more than a hundred countries, including the United States, the country that hosted his dreams.

The education minister had just created the first system of educational credit that allowed young people from all over the country to study abroad. While Yolanda Pulecio was working for the most disadvantaged, Gabriel Betancourt was busy with what was to be his life's greatest work: giving nonwealthy Colombians the same education opportunities that only a lucky few could afford until then.

My parents married in the late 1950s. Astrid was born in 1960, and I the following year. Just after my birth, we left Bogotá to spend several months in Washington, D.C., where my father joined a team President John F. Kennedy had set up to launch the Alliance for Progress, which was to promote development in Latin America. He was named head of the educational commission. When Kennedy was assassinated, this project came to an end, and my father was deeply saddened by it. But immediately afterward he was appointed to the post at UNESCO, and we went to live in Neuilly. I remember my parents as extremely busy, but determined to escape the whirlwind of their activities in order to find time to take us on their laps, answer our questions, read us a story.

My father listened, smiled, took the time to explain, but did not play with us. "I'm much too old to play, but I can read you a book. Choose a book." He was big and strong, his forehead wide, his brown hair slicked back. He wore heavy, horn-rimmed glasses. But, the severity of his face immediately disappeared when he smiled. Papa's smile! All the benevolence in the world shining down on our little heads. My mother was quite willing to play. She was spontaneous, sensitive, active; she has the elegant simplicity of Audrey Hepburn, yet sometimes she reminds me also of Sophia Loren, with her warm Italian beauty. In Mama, there was sunlight, appetite, warmth to spare; she couldn't hide her Italian origins.

To follow her husband, she must leave the *Albergue* in the hands of the team she has put together in Bogotá, but she takes advantage of her years in Paris to study the French system of aid to children. She talks to a great many people, and my father's position helps open doors for her. At this time, France is dealing with a massive influx of *pieds-noirs*, French colonials born in Algeria who were forced to leave the country after independence, and my mother sees in this a similarity to the arrival in Bogotá of poor peasant families driven out of the countryside by poverty or violence or both. It is the children of these families that she has taken off the streets, half-starved. How does France manage to integrate its *pieds-noirs*, to house them, educate them, make jobs for them, subsidize them? My mother listens, observes, takes thousands of notes, and draws up plans, while waiting to return to make the *Albergue* what it is today: the best known children's aid organization in the Colombian capital.

In 1966, the year I turn five, we go back to Bogotá. We return because the newly elected president of Colombia, Carlos Lleras, wants to make my father minister of education, and thus he becomes a minister for the second time, with the same portfolio. Astrid and I discover Colombia, which we don't remember at all, and since we speak French as well as Spanish we're enrolled in the same French school that our own children will attend twenty-five years later. My mother, who has just turned thirty, also begins a new life: she goes into politics.

She chooses the position best suited to help her work with children—assistant for social affairs to Bogotá's mayor. She's one of the first women to hold a responsible post in the capital city's administration. This enhances her image, but Colombians have too often been deceived by their politicians to take them seriously. So they wait to see if this well-off young woman, known for her beauty and her big heart, will now take advantage of her power to enrich herself, as do most politicians in our country, instead of pursuing her initial goal. Soon my mother becomes a leading figure in the creation of the Institute of Social Welfare. She puts into this project all her expertise, skills, and experience. She comes back from France with a "model" for a system that addresses these kinds of issues. Prior to this, solving social problems in Colombia was a matter of charity. With the creation of the ISW, it becomes a matter of state policy. For Colombia, which has never taken a serious interest in its downtrodden, it is a revolution—such a revolution, in fact, that the wife of President Lleras, seeing the advantage she can draw from it, quickly sets about appropriating this great innovation. Mama doesn't care. Unlike my father, who sees to it that his name remains attached to the reforms he has initiated, she regularly lets her ideas be stolen. Colombians know this, and in a way they love her for it even more.

In the late 1960s, Yolanda Pulecio's fame increases, and confidence in her is growing, especially among the poorest people. It's conceivable that she will soon hold an important office at the national level. The path my father is following is more complex: after having come very close to the top echelon of national politics, he's about to fall suddenly into disgrace. This fall will slowly undermine my parents' relationship and eventually lead to their separation.

Since the ministry of education enjoys a reputation for exceptional integrity in Colombia, my father is soon approached by a group of businessmen and young technocrats trained in the United States. They see in him the man who could lead Colombia out of its ancestral corruption into a community of the great democracies. They ask him to become a presidential candidate. He thinks about it, then refuses. It's not the right moment, he says. But my mother urges him to accept. She thinks this is precisely the right moment; the Colombian people urgently need a man like him. The survival of the most disadvantaged and the future of children yet unborn is at stake, and he doesn't have the right to evade his responsibility.

While this personal debate is going on, my father continues to conduct his ministry in his own way: curt, technical, rejecting compromises, half-measures, and anything that looks like a payback. In a country where every minister distributes posts in his administration to his political associates in exchange for the votes of their whole families in the next elections, my father refuses to participate in the government unless he can choose all his staff on the basis of their technical competence instead of their electoral weight. He closes his door to all kinds of solicitors, and worse yet, demands that legislators who want to see him justify their request in writing—an incomparably effective technique for discouraging dirty dealing. All this ends up making him unpopular, and at the end of 1968, yielding to a political class no longer willing to put up with such an absurdly inflexible minister, President Lleras thanks him by naming him Colombia's ambassador to UNESCO, a polite way of exiling a man who has caused too much trouble.

So in January 1968 we leave for Paris again. But this time it's difficult for my mother. She has to abandon everything she's begun at Bogotá's city hall—and for what? To accompany a man who has deeply disappointed her by refusing to seek the presidency? Even if she doesn't put it this way, even if it will be years before she understands it, at that moment she feels that my father has given up the great battle she dreamed of. It's clear to her that this prestigious post of ambassador is a gilded retreat. My mother is only thirty-three, and retreat seems to her all the more horrible because Colombia's needs are immense, and she knows she's capable of meeting some of those needs, especially with regard to children.

We're back in Paris, in an enormous apartment on Avenue Foch. It is decorated with great taste: signed eighteenth-century furniture, paintings by old masters—I recall in particular Dürer's *St. Jerome*, which frightened us at night—Chinese bibelots, carpets, a hanging garden. My parents lead a busy social life. They're received in all the high places of government (Georges Pompidou has just become president of France), and once a week they throw parties for two or three hundred people. In this whirlwind, they no longer have much time to be concerned with the details of our everyday life, and so they've hired Anita, a Portuguese nanny who has lived through all the convulsions of the century. With this intelligent and infinitely loving old lady I am to have my first "philosophical" conversations: "You must not forget, Ingrid, that the world does not resemble the one you're living in today. Reality is painful, life is difficult, and someday it may be painful and difficult for you too. You must know this and prepare yourself for it."

I'm ten years old, but my memory is filled with painful and violent images of Colombia. My parents used to tell us about the "Violence"—the historical name for the period between 1948 and 1965—that they witnessed in their homeland, my mother in the Tolima and Santander regions, my father in the Antioquia region. Armed men would come to set the villages on fire, rape women, and kill men and children because they were "the seeds of evil." This happened everywhere in Colombia, in a war between liberals and conservatives. In regions controlled by liberals, all conservative citizens were exiled, forced to leave behind everything they had, or they would be killed. In the regions controlled by conservatives, the same thing happened to liberals. Party affiliation was akin to religious affiliation. You were born either liberal or conservative, depending on your family's tradition. And you grew up feeling hate and intolerance against the other party, because they had threatened or killed your loved ones. I also have recollections of the hungry, forgotten children my mother saved in Bogotá, so I understand what Anita is telling me. And I love her for it, because it proves that for her I'm more than just a privileged child (have I said that my first communion was administered by Pope Paul VI during his trip to Colombia in August 1968?). This conversation proves that Anita takes me seriously and believes in me.

In the attic of our luxurious building lives Monsieur Constantin with his little dog, Pat. My parents hire Monsieur Constantin to help with every reception. Because I love animals, Pat's arrival is a special gift. Soon our mutual interest in this dog forms a bond between Monsieur Constantin and me. We become friends, and the life of this old Russian aristocrat perfectly illustrates what Anita has been saying to me in her kind, grandmotherly way. Formerly powerful and respected, Monsieur Constantin had to flee Russia after the Bolshevik revolution. His entire family and fortune were wiped out. Having lost everything, he is reduced to serving petits-fours to people who treat him like a servant. Nevertheless, his culture is immense. I have a deep love for this modest, refined man.

Astrid and I go to school at the Institut de l'Assomption in the rue de Lubeck, right in the middle of the fashionable sixteenth arrondissement, along with many well-off French children. We take the no. 82 bus, which passes in front of our building and lets us off in front of the school. Except on the rare mornings when the chauffeur of the Colombian real estate magnate Fernando Mazuera, a friend of our parents who lives across the street from us, agrees to take us—in his Rolls Royce. On those days, it amused us to show off this temporary luxury. But fortunately, Anita keeps an eye on us. We will never be completely taken in by the glitter of gold, by mere appearances.

Our parents also keep an eye on us, and in our everyday lives there are constant reminders that beneath the splendor that surrounds them looms an inescapable reality. Many Colombian political figures come to our house for private dinners; they include the former president of Colombia, Carlos Lleras, whom my father continued to deeply admire, understanding the reasons that Lleras removed him from the ministry of education.

I also recall others: Misael Pastrana, the president of Colombia at the time and the father of Andrés Pastrana, who would be elected president in 1998. The painter Botero, father of the future minister of defense, Fernando Botero, with whom I was to have bitter clashes and who would end up in prison twenty-five years later. Virgilio Barco, another future president of Colombia, the writer Miguel Ángel Asturias, and many others sat around our lush dining table. All these intelligent people seem terribly concerned about Colombia's future. One evening I stay up listening to them, and I am so upset by what I hear that once I have been sent to bed, I get up again and go to hide under the grand piano, in a corner of the living room, in order to follow the conversation. I am upset, as I realize later on, because of my tendency to take literally the words uttered by adults. They say that the election of a certain Turbay (which took place a few years later) will be "a catastrophe" for the country, that this or that economic decision will surely lead Colombia into "an unprecedented disaster," and I see my country sinking, people dying. I often return to my hiding place under the piano, and sometimes emerge with my temples burning, my stomach in knots, ready to burst into tears—so awful, truly terrifying, do I find my country's fate. Today, I believe that my political vocation was born under this grand piano at the beginning of the 1970s.

Of all these guests, the only one with whom I form a relationship of an exceptional tenderness is Pablo Neruda. He spends more time in Paris than he does in Chile, and our door is always open for him. Moreover, he often comes unannounced. Few adults know how to find the right words to share a feeling with a child, but Pablo knows how; he has that special gift. I'm aware that he's a poet, though I don't know how famous he is, and I have no idea that he has just received the Nobel Prize. One day I say to him:

"You know, I write poetry, too."

"Really? Well, let's exchange poems, okay? The next time you'll recite one of yours, and I'll recite one of mine."

This becomes a ritual between us. As soon as he comes in, I run toward him, he hugs me, and then we exchange the best lines we've written. At least I do. "Ella es mi colega" ("She's my buddy"), he tells my father. I still have this note he wrote me: "Ingrid, I'm leaving you a flower. Your uncle, Pablo Neruda." He died in Santiago in 1973.


My mother is at her husband's side, the mistress of a splendid household, attentive to everyone and to the slightest detail. Yet she's bored. Her heart is in Bogotá, and what happens here in Paris seems to her unimportant and superficial in comparison to the news she receives from her teams in Colombia about the increasingly distressing plight of the children. Often she betrays her concern. Instead of telling us about the evening she has spent at the French foreign ministry or at the theater in Paris, Mama tells us, in great detail, how in Colombia they've saved a five-year-old boy who was eating out of a restaurant's garbage cans.

We sense that she's just waiting to return to Colombia. Astrid and I show our joy when my father tells us we're going home. We've spent five years in France, and we've just returned from a year in boarding school in Sidmouth, in southern England, where we learned English. To ease the transition, our parents decide to travel by boat rather than by airplane. We embark at Genoa for a month-long voyage. Suddenly Papa, who hasn't had a spare minute for a long time, relaxes, free of any obligation. For me, it's a giddy delight to have him available, rested, attentive, benevolent. For hours, we read together—the *Gulag Archipelago*, I remember—and above all we talk as we've never talked before—about France, about Colombia, about everything that remains to be done at home to attain a fair balance of democracy, ethics, and respect for others. He tells me something I've never ceased to think about: "You know, Ingrid, Colombia has given us a great deal. It's thanks to Colombia that you have come to know Europe, that you've gone to the best schools and lived in a cultural luxury no young Colombian will ever experience. Because you've had so many opportunities, you now have a debt to Colombia. Don't forget that." Fifteen years later, I will repeat these words to myself when I suddenly break with my gilded life as a wife and mother in Los Angeles to return and pay my debt to Colombia.

I'm thirteen years old, and I can't know that on this ocean liner we're living our final moments of family harmony. As soon as we arrive in Bogotá, our parents buy a very fine house overlooking the city, only ten minutes from the French school where we will continue our studies. Everything seems to be going well. But our parents are secretly moving away from each other. Papa is very busy, constantly traveling for UNESCO, taking part in this or that international meeting; he's always been deeply interested in cultural exchanges. But Mama no longer goes with him. She doesn't want to be his smiling, devoted traveling companion; she wants to live by herself, to return to her social work in Bogotá. She thinks she's probably already sacrificed too much to this diplomatic life that's so far removed from the reality of poverty. My father, always between one airplane and the next, does not listen to her, does not understand her.

One day, Mama goes away. Only then does Papa realize how much he cares for her. Did she say she was leaving forever? No, we will find out later that she mainly wanted to be alone in order to think. But for him, it's a devastating blow, and he decides to respond with a still more radical blow, probably to avoid collapsing himself.

It's a Saturday morning. Astrid and I are with him; I'm now four-teen and she fifteen.

"I'm going to work at home today," Papa tells us. "I'll drop you at Los Lagartos Country Club and I'll come back to pick you up in the late afternoon."

We sense that he's closed up, tense. He goes away, and we spend a gloomy day together. What's happening to our parents? For a long time, their happiness had seemed to us clear, contagious, luminous, and now they're both behaving as if they were autistic.

Around six, Papa returns to pick us up. He's very pale and looks exhausted.

"Astrid and Ingrid, listen to me. I've just sold the house; your mother has left. Our life together has come to an end. You're going to live with your grandparents for a while."

"You sold the house? In one day? That's impossible, Papa! You haven't done that!"

"Yes, I have. I sold everything. Everything."

"You're crazy, Papa. You're not serious, you can't have sold every-thing in the house in one afternoon. What about my dogs? You didn't sell my dogs?"

"No, your dogs are the only things left. We'll go get them right away."

He goes in first, opens the door to the house. We're in shock. I can't think of words to express the silent desolation of these empty rooms where, this very morning, despite Mama's departure, we could still believe that our family life was eternal. Only the marks left on the walls from our pictures show that we once lived in this house. Father has in fact sold everything, not only the furniture we brought back from France, but also our little beds, our books, our trinkets—every-thing that helps to shape memory, to fight against time. It's his way of wiping out our past existence, denying it, erasing Mama and the bonds that unite the four of us. For Astrid and me, it's an absolute, irreparable disaster. From now on, there'll always be our life before the family household was destroyed and our life afterward. We'll never recover from this crushing act, and we'll never be able to mention it without feeling the full force of its pain.

If my mother was still dreaming that we'd all live together again, that's now out of the question. By destroying everything, my father has hastened the onset of his own unhappiness and begun a war that will last ten years.

Divorce proceedings are quickly begun, and Papa sets the hostile tone by demanding custody of us and forbidding us to see Mama—a prohibition we don't respect, of course. This divorce becomes a nightmare for me and my sister, even more so than it does for Mama. My parents are too well known for the newspapers to ignore what's happening, and, in 1970s Bogotá, people of high standing don't get divorced; it's simply not done. The newspapers glorify my father, the former minister, the former ambassador, and cast opprobrium on a wife "who is abandoning a man to whom she owes everything." My mother is "scandalous," and all the women who envy her beauty take advantage of this to criticize her frivolity, her pride, her selfishness— all characteristics entirely foreign to her. Thus a woman who is separating from her husband in order to resume an active role in society finds herself ostracized by that same society, pointed at, slandered, condemned. The height of cruelty comes when the court deprives her—a woman who has led the fight for disadvantaged children—of custody of her own children.

Astrid and I are revolted by the court's decision. Though we love our father deeply, we have to admit that he was very busy during our childhood, whereas Mama was constantly present. Taking us away from her is unjust and disgusting. I don't fail to say this to Papa, and I get my first slap for having done so.

"Ingrid," he says to me one day, "I remind you that you're not to see your mother. She can't help but have a bad influence on you.

Besides, read what the newspapers say about her. I'm not making these things up."

"I don't give a damn what the newspaper says or what you think about her!"

Wham! He slaps me. Poor Papa!

This is a dark, hopeless time for Astrid and me. Our father, so solid in life, now seems to us a wounded, bitter, gloomy man. It's difficult to pardon him without understanding the depth of his suffering. Our mother, on the other hand, is also suffering greatly, but silently, and this is even worse. In order to see us, in defiance of the prohibition, she has rented a small apartment with a window looking out on the playground of the French school. At every recess I stand where she can see me, and we send each other kisses and other signs of love. On some evenings, we wait until Papa goes out and then, in our nightgrowns, we run from our new lodgings near the school over to Mama's apartment building.

Then Mama does something that amazes me and gives me the finest possible lesson in courage. Persecuted by the press, vilified by the whole upper class, deprived of her maternal authority, she dares to seek reelection to the city council. Without a dime, without anyone to support her, she begins her campaign—alone. With her slogan, "Let me work for your children," she thumbs her nose at those who accuse her of being an unworthy mother. She's magnificent! Moved and enthusiastic, I devote all my free time to her campaign. I put up her posters, I distribute her pamphlets, I accompany her to public meetings. She's waging this battle with a heart full of rage. She wants above all to prove to herself that she's not what people say, but just the opposite.

In the wealthy neighborhoods in the northern part of the city, her old friends, even those who were her guests in Paris, close their doors to her, but in the southern part where the families whose children she has helped live, she's given an incredibly warm welcome. She has had to sell her jewelry; she no longer has anything. One evening, two men whom she'd cared for when they were children come to her apartment, their arms filled with food: "This is to fill your refrigerator, Mama Yolanda." And Yolanda laughs to keep the tears from coming.

Three months later she's elected—by the votes in the southern part of town. I tell myself for the first time that every battle for justice is ultimately successful, and one never fights in vain.

However, she won't be on Bogotá's city council for long. People turn their backs on her; they constantly remind her that she's a divorced woman and therefore unworthy of her office. Over the months that follow, I see her collapsing, losing heart. I remember my adolescent anger as I try to cheer her up: "But Mama, you shouldn't care what those bastards say! So what if people talk about you, that proves that you're making them uncomfortable, all of them. They're jealous, they want to see you destroyed, they can't stand seeing you go on fighting. I admire you, Mama, and that's what ought to count."

It's not enough. The year I turn sixteen, my mother accepts with relief an offer to go back to Paris to work at the Colombian embassy. She packs her bags and leaves her country. She won't return for another decade.

For me, Mama's departure is an additional heartbreak. Coming from Assomption, that orderly Parisian school, I've had a hard time readapting to the crude ways of Bogotá's French school. During the first weeks, I spend most of the recess hiding out in the rest rooms to escape my schoolmates' badgering. Then I become hardened, and the pitiless war my parents are still waging against each other, which the whole school knows about because it's reported in the press, makes me a rebellious, combative adolescent, even more stubborn than most. Papa, with whom I am now living, bears the brunt of my rebellion. Our relationship is tense, conflictual. "You know," he tells me in one of the rare moments when we're able to laugh at ourselves, "you've given me all my white hair."

Astrid, who's eighteen and thus no longer subject to the custody law, has gone to France to live with our mother. I've been asking permission to spend a month in the summer with her, but Papa's ignoring me.

One day I burst into his office.

"Papa, with or without your consent, I'm going to see Mama. That's how it is. So get me a plane ticket, please."

He looks up, remains silent for a moment. Then he says coolly: "You'll have your ticket, Ingrid. But you won't have my consent. If you really want to leave, you'll have to get permission from the judge who awarded your custody to me."

"Okay. Give me his name and address and I'll go see him right away."

He's visibly shaken, but he gets up, searches among his papers, and gives me what I've asked for.

The judge's office is in the southern part of Bogotá, so I have to cross the whole city. In Colombia, a young woman does not travel long distances alone, especially not in dangerous neighborhoods like that one. Papa, of course, is not going to lower himself by offering to accompany me. Without a word, with no sign of his concern, he watches me get ready.

I take the bus, I get lost, and of course, in the crowd, the small amount of money I have on me is stolen. Finally, I arrive at the right address: a gloomy building stinking of urine. Slumping, weary people are waiting everywhere. I ask for my judge and am told how to get to his office, at the end of a dark corridor. I sit down with the other people waiting to see him. Here, everything is dirty and discouraging. Finally, the judge receives me. He's a bald man with a rather kind face. His eyes betray his fatigue, weariness.

I'm wound up like a clock.

"After all, it's weird that the Colombian judicial system would force a teenager to go all the way across Bogotá, taking the risk of being mugged, just to get permission to go hug her mother. Do you realize what sort of society we're living in? And you, you're a judge, and you condone this shit! You think that's okay? You find it absolutely normal that I have to spend two hours in a bus to come here and beg you . . ."

He lets me spill my guts, dragging the whole country through the mud and him along with it. When I finally fall silent, he says,

"Well, what do you want, miss? A document signed by me that will allow you to go hug your mother? Fine, I'll give you that document right now. Then you show it to your Papa. You see, it's not so complicated. In any case, it's not worth getting yourself in such a state."

I give the letter to my father. He laughs as if he finds it all very funny, and only then do I realize that I didn't really need to ask the judge.

"Your papers are in order, my dear. When do you want to leave?"

I'm back in France. It's the middle of the summer and I'm so happy! I didn't let Mama know I was coming. I get a taxi at the airport and go directly to the Colombian embassy.

"Madame Yolanda Pulecio, please."

"Do you have an appointment?"

"I'm her daughter."

"Oh, excuse me, miss. Go ahead, it's the third office on the left."

The door's open; no one's there. I go in and hide behind the open door.

My mother appears. I see her stride quickly across her office, a bunch of files under her arm.

Then I slam the door. She turns around, sees me, and bursts into tears.

"Mama . . ."

She lives in a decent apartment on the Boulevard St.-Germain. But it's no longer our place on Avenue Foch. Mama works; she's no longer an ambassador's wife. But she has bloomed in a profound way. Of the people who used to come to our apartment on Avenue Foch, there remains a small circle of faithful friends, among them Gabriel García Marquez, the Colombian Nobel Prize–winning author, and his wife, Mercedes, of whom Mama is particularly fond, as well as the painter Fernando Botero. I stick close to her throughout this month in Paris; we caress each other, we make up for lost time, I count the days. A month is so short in comparison with the long year of separation that awaits us. I'm in my last year of high school.

The woman I am today is born during the year I spend in Bogotá, after my return from Paris, preparing for my high school baccalaureate examination. It is the best part of my life up to this point, a year of inexhaustible intellectual and sensual discoveries, a year of learning to be free. I begin to study philosophy, and my passion for literature is confirmed by it. With the support of the school's administration, I stage Albert Camus's play *Le malentendu*, and find myself attracted to theater. I stay up all night, talking endlessly over a bottle of wine in the back room of a smoky bar—I who have been so prudent and reasonable for such a long time. And I have my first love affair.

During this year, which is so eventful, so transforming, I refuse to cheat, to lie to myself. I decide that the freedom to go astray, in particular, is inseparable from the necessity of assuming responsibility for one's acts, whatever they might be. So I vow to tell the truth, in particular to my father. I tell him everything—about how I'm living, about staying up all night, about returning at dawn to prepare feverishly for an exam set for two hours later, and also about my astonished delight at the feelings of the heart, of love. Everything serious and forbidden I do during this year and I hide nothing from Papa. I know it's hard for him to hear about it, and all the harder because he's almost sixty, but I am trying to make him a part of my life. I want to preserve this contact at any price, and the hell with his white hair.

When I tell him that I've made love with a boy of my own age, it's a terrible blow for him. I see him go pale. For him, it's unthinkable that a girl of seventeen, without being married . . . But I insist that he listen to me, that he tell me what he thinks, and even that he counsel me—why not? I've accepted my parents' life, their breakup, and I want him to accept what happens to me as well. But he can't do it right away, and he immures himself in a terrifying silence. For weeks he won't speak to me. We share our meals without exchanging a word or a look. I tell myself: If he wants to punish himself, okay. He doesn't want to talk to me? Well, I won't talk to him, either!"

Then one day my boyfriend's sister tells me she's getting married and invites me to the reception. But I don't have a suitable dress; I've nothing to wear. So I write a note to my father, since we no longer communicate any other way: "Papa, I've been invited to Mauricio M.'s sister's wedding. I don't have anything to wear. Ingrid." He reads the note, and his face suddenly lights up.

"Well, let's go buy you a dress, my dear."

He's come back to life, and so have I. He has traveled in silence the long path leading back to me, and here we are together again, accomplices. He has me try on one dress after another. With Mama, he'd never allowed himself the pleasure of dressing a woman he loved; now he discovers it, and his eyes fill with tenderness.

"Which one do you prefer?"

"I'm not sure. I like this one because it's black and black's good on me. The white one's great, too, but the long dress is the most elegant for a wedding. Don't you think so?"

"I do. You know what? We're going to take all three of them!"

Papa has come back to me, at last! I can talk to him about Mauricio. He listens to me with his customary intelligence and benevolence.

Soon, even before this last year in high school is over, Mauricio asks me to marry him. I'm well aware that we're both children, but I also feel that I love him. But would I marry him? I don't know. On the other hand, I think the news will please my father; if Mauricio asks me to marry him, that proves that he really loves me, sincerely, deeply, and since Papa thinks that love cannot, and should not, bloom except within marriage . . .

Papa listens to me, and he's marvelous, even better than I could have imagined—infinitely respectful, liberating.

"You know, Ingrid, this is a decision you'll have to make by yourself. All alone. If you want to marry this boy, marry him. But if you don't want to, don't. I have no role to play in your choice. If you say yes, you're the one who's going to have to live with him, you alone. Think carefully about what you want to do with your life, and make up your mind. But whatever you decide, one way or the other, it'll be all right with me."

On that day, Papa gives me wings. A few weeks later, I break off my relationship with Mauricio and pack my bags to go to France.

# CHAPTER THREE

IN 1980, I turn eighteen. I've been awarded my baccalaureate. I live in Paris and I'm preparing for the examination for admission to Sciences-Po.*

I've become closer to my mother, who's still working at the embassy, and thus farther from my father, who's now living alone in Bogotá. Farther only in a geographical sense, however, because after the difficult years of adolescence we have become very close. Papa has overcome the suffering caused by the divorce, and he's putting all his ability to listen and all his love in the service of my future.

We discussed for a long time my decision to enter Sciences-Po. My father is ambivalent about politics. He thinks there's nothing more noble than to serve one's country, as he himself has done as head of the ministry of education, but he has the deepest scorn for professional politicians who pursue their careers at the expense of the state and, naturally, for the Colombian political class that is pillaging the public treasury. He sees me more as a philosopher far from the battlefield, and can't imagine that someday I'll be part of the corrupt

---

*"Sciences-Po" is the name commonly used to refer to the École de Sciences Politiques, the premier school in the field of political science in Paris.

oligarchy that he refused to have anything to do with when he was minister. To tell the truth, I don't think about it, either. Not for a second do I imagine that fifteen years later I'll be elected as a representative, then as a senator, on platforms devoted entirely to the battle against corruption. No, but I recall the intense feelings I had when, hidden under the piano in our apartment on the Avenue Foch, I listened to some of the most prominent Colombian politicians describe the risks our country was running if this or that decision was not made in time. The desire to influence these decisions, to affect the destiny of the country, is certainly already there, buried somewhere in my unconscious, but I don't know how to express it in words, and I try to find a way of convincing my father that Sciences-Po is precisely the right school to help me realize my deepest ambitions. Suddenly I remember our Sunday mornings in bed, on the Avenue Foch. Papa would read through the Colombian newspapers, which he received once a week, while I, squeezed in between him and Mama, read the headlines. If something felt worrsome to me, or if he laughed at a cartoon, I demanded an explanation. This sometimes annoyed him, but he tried to answer my questions.

"Do you remember how much I loved to read the newspaper with you when I was little?"

"Of course I remember. That amused your mother very much."

"I was already very interested in the news, and that's why I want to go to Sciences-Po. I like philosophy, but I want to live in the present, I want to act."

In the meantime, I spend hours in the library, far from the center of action. I have an insatiable thirst for knowledge. I want to understand how institutions work, how they combine the executive and the legislative functions, and behind all that, of course, I want to discern the perversities, imagine the guardrails. Why do certain democracies, such as France, succeed in keeping themselves firmly away from corruption, whereas others, such as Colombia, sink into it body and soul? I love the library at Sciences-Po, the respect for thought and for silence one finds there.

At this time my desire to immerse myself in study is so strong that I decide to live alone rather than with my mother. Once again my father comes to my aid: "Find yourself an apartment, Ingrid, and don't worry about the rest. I'll take care of it."

One day while I'm living this solitary and gilded life, a child of four comes up to me in a restaurant. He's adorable, a little angel, and we exchange a few words, but when I instinctively try to catch the eye of his mother at a neighboring table, I encounter the smile of a man. He's alone with his son, and he's about thirty years old. We get on well; is the child's mother traveling? No, he's divorced. By the way, he's looking for a baby-sitter. That's convenient; I need money. We laugh.

Fabrice and Sebastien have just come into my life. Fabrice has recently begun working at the ministry of foreign affairs, as a commercial attaché. We share an interest in politics, a curiosity about what's happening outside our borders. He's French, and he thinks of Colombia as a violent country in turmoil, but I don't seem to correspond to this image. He thought I was French, with my auburn hair, my pearl necklace, and my impeccable command of the language. I tell him about my strong, sentimental connection with France, but I also tell him how much I love Colombia. I cannot guess that ten years later, my imperious desire to return to Bogotá will cause our love affair, which is now about to begin, to literally explode.

Fabrice is intelligent and cultivated, open to the world, elegant, very handsome. In short, he has all the attributes of the masculine ideal transmitted to me by my father. Very soon, we both become sure of our love, and want to commit to each other forever. Then there's Sebastien, linking us like a ray of happiness. I find myself mothering him, and discover that I want a family. We get married, we travel, we have other children. When we're together all our dreams seem possible. We're amazed, confident. Infinitely confident.

Fabrice is soon posted to Montreal. The pain of his departure is assuaged in part by the pleasure I feel in finding myself alone with my books. I go to Quebec for short visits, and as soon as I get back to Paris, I hole up by myself. I have enrolled in Sciences-Po, and the farther I advance in my studies, the more I feel in harmony with the complex mechanisms of government. Public affairs interest me deeply, and now I understand perfectly how the formidable mechanism of government operates, but also how fragile democracies are, their strict dependence on the personal ethics of each elected representative, each government official. I dream of putting all this into practice, but at the same time I'm in no hurry to begin. My priority, once I've taken my degree, is finally to live with Fabrice full time. We get married.

In 1983, our first year of life together, he's posted to Quito, Ecuador. For me this posting is a real gift. Now that we are going to live in a country bordering on Colombia, I express the desire to return to live in my own country. We talk about it often, but Fabrice is not enthusiastic about the idea. Colombia frightens him, somehow. But he's quite willing to learn Spanish, and so is Sebastien. Soon they both speak it perfectly, especially Sebastien, without accent. For me, Quito is a first step toward my return to Colombia.


At least, I hope it is, but exactly the opposite happens. These three years spent in Ecuador dissuade Fabrice from asking to be posted to Bogotá. From a close perspective, the spectacle of Colombia is too discouraging, and unfortunately it corresponds to his worst fears. The economy is stagnating, and while the drug traffickers declare open war on Colombia's institutions (the assassination of the minister of justice in 1984 is the first shot), the guerrillas decide to resume armed conflict. Everything suggests that the country is on the brink of disaster (in 1989, twenty-three thousand people will die), and Fabrice refuses to consider raising our children in this whirlwind of terror.

And I've just gotten pregnant. Melanie is born in September 1985. Her first steps are taken not in Bogotá, as I'd hoped, but in the Seychelles, under the luminous sky of the Indian Ocean. Fabrice has just been posted to this heavenly archipelago.

The extraordinary happiness of having Melanie, of becoming a mother, suddenly revives the memories of my childhood suffering after my parents' divorce. I feel nostalgic for a vanished family happiness and find myself dreaming about a reconciliation, about a family once again united around my little girl. December 25 will be her first Christmas; the same day I will celebrate my twenty-fourth birthday, and December 31 is Mama's birthday. All this inspires in me a diabolical plot whose risks I do not gauge, so much do I want it to succeed.

I write a long letter to my mother, inviting her to spend Christmas with us—for her granddaughter and for me, because I feel so uprooted. I ask her not to tell anyone in order to keep Papa from finding out, because he would be hurt if he found out that she's coming first and that he had not been invited. Then I send exactly the same letter to my father, begging him to join us and also urging him to conceal his trip so that Mama doesn't find out, for she would certainly be very hurt, etc. A few days later, both of them write to say that they're delighted to accept my invitation and promise to keep it a secret.

They've been fighting for precisely ten years; how are they going to be able to stay under the same roof? As far as harmony goes, the situation might very well turn into a tragedy and Melanie's first Christmas into a debacle.

Papa sets out a week before Mama. I've told Fabrice, his relatives, everyone: Papa mustn't find out about Mama, because otherwise he's quite capable of leaving on the first available plane. But the drama explodes the day before Mama arrives. Who betrayed the secret? I'll never know.

"Ingrid, you had no right to do this to me! If you'd told me, I'd never have come."

"You're right, Papa, I should've told you both, but if I had you wouldn't be here, and I wanted to be with you both. It was very selfish on my part. But now, if you really want to do something nice for me, spend Christmas with us and leave immediately afterward if you want. You'll be staying at opposite ends of the house. If you don't want to talk to each other, you won't have to. The other solution is that you leave on the same plane Mama comes in on; that way you won't see each other at all. That would cause me a great deal of pain, but I'd understand. I went too far, forgive me."

Papa sulks. I think he's actually going to get on the plane, but at seven the next morning, as I'm getting ready to go to the airport, he's still sound asleep and his bag is not yet packed.

"Mama, Papa's here, at my house."

"Really? That's splendid! And does he know I'm coming?"

"Yes."

"Is he willing to see me? What did he say to you?"

"You'll find out."

In fact, had I not seen it with my own eyes, I wouldn't have believed it. They spend a month outside time, outside the world, talking, patiently rebuilding many of the bridges they've burned, forgiving each other, laughing, crying. And I witness all this, my heart constantly wrung with emotion, I'm so happy for them and also for Melanie, who at this point knows nothing about her grandparents except this incredible bond between them. Years later, I hear her ask my father:

"Why didn't you and Mama Landa live together?"

"Because with all these books, my dear, I no longer had room for your grandmother."

A few months before this unforgettable Christmas in the Seychelles, Mama left her job at the embassy and returned to Bogotá for good. She has recovered confidence in herself, regained her strength, and in early 1986 she begins, with an incredible energy, her campaign for a seat in the House of Representatives. She wants to be a representative in order to resume the social work she most cares about, to speak on behalf of peasant families who've been driven from the countryside to the cities by the guerrillas and the drug traffickers, families whose children wander around Bogotá before they're picked up by humanitarian organizations.

My mother now becomes my chief connection with Colombia. We don't let a day go by without speaking on the telephone. She's on the scene, remarkably well informed, and she learns even more after she finally succeeds in being elected to the legislature by campaigning for the enactment of a protection code for children. Everything she tells me upsets me. Colombia seems doomed to misfortune, to destruction. If it's not nature striking the blow—as at Armero, where the eruption of the volcano Nevado del Ruiz buried twenty-five thousand people—it's the guerrillas, who are attacking the very heart of the state. In this year, 1985, M19, one of the most active guerrilla movements, besieges the Palace of Justice, which is the seat of the Supreme Court. When the army finally recaptures the building, more than a hundred people are dead, including half the judges of the Supreme Court.

While my country suffers, while my mother is fighting, I'm in the Seychelles, in a tourist paradise. I'm the wife of a French diplomat, living in a splendid house, with nothing to do except take Melanie for walks and give orders for the dinners and receptions we organize from time to time. I feel out of place. My happiness seems more and more meaningless, even indecent, because it has so little to do with my own people. All the happiness in the world would seem ridiculous to me, in comparison with what's going on in Colombia. But what can I do? Fabrice is happy, and since he's not Colombian, why should he take any part in the tragedy going on there? Years later, I will remember Mama on the Avenue Foch—the brilliant, devoted wife of the Colombian ambassador, secretly torn between her love for her husband and her despair at not being in Bogotá, alongside those who count for her—and I will say to myself: truly, history sometimes repeats itself in bitter ways.

# CHAPTER FOUR

IN THE SUMMER of 1986, I am completely anxious: I decide to spend two months in Colombia with Melanie, on the pretext of showing her the country. Fabrice is busy with his work, and the two of us fly off alone. It's been more than seven years since I left Bogotá to enter Sciences-Po, and I miss everything that emanates from that unbelievable city. The austerity of the mountains (Bogotá rises wildly, furiously, at an altitude of 8,500 feet); the mad bustle of its streets; its skies, which are often leaden; its devastating rains; and, always, the dark melancholy in the eyes of its citizens. What am I expecting from this trip? Nothing. Everything. That Colombians will recognize that I'm one of them, that they'll adopt Melanie, that they'll let us breathe the same air they breathe. I'm not living in reality; I'm too full of ideals fed by distance and guilt, too full of a stubborn, naïve love for a country whose suffering I've never shared. But everything that happens to me during these weeks will help throw me into a cruel reality.

Mama is there—vibrant, involved in a dozen different projects, constantly running back and forth between her legislative office and her practical efforts. She discerns my excitement mixed with doubt and confusion as well, and she suggests that I accompany her on a trip to the Atlantic coast, more precisely to the Maicao region, in the north-west. It's an area where the people make their living by smuggling, a zone where laws are not enforced, where some people are getting enormously rich while many others are being killed, where the common people are mired in poverty. Mama tells me she's going, with about twenty other legislators, to meet these people, listen to them, and try to find solutions to their problems. I believe her, and we both believe that this is the best way to proceed. So we leave together.

It turns out to be an astonishing journey on which we work, listen, learn, laugh a lot, drink a lot. We're surprised by the unusual turn of events; only later do we realize how inappropriate the trip was. Everywhere we go we're welcomed warmly, with banquets, exhibits of local folklore, speeches. Where are the common people? It's the local elected officials who do the talking, and these men go all out for us, obviously more concerned with promoting themselves than with expressing the problems of the villagers, families, small businessmen. One member of our delegation is having a field day, being more praised and courted than the others. His name is Ernesto Samper. He's a friend of my mother's, also a representative, and endowed with a sense of humor no one can resist. He's uninhibited, missing no opportunity to ridicule someone or other, making fun of everything, and he enjoys a drink. Sometimes he pretends to be paying careful attention to what a petitioner is saying, but then he forgets and leaves the paper on the table.

During a dinner with the big honchos of the smuggling community, I'm astounded to hear Samper talk to them in a demagogic manner, as if he were campaigning for election: "You're living off a trade that evades taxes, of course, but is there anyone in Colombia who doesn't benefit from it? We have to take steps, but steps that apply to everyone. In the meantime, I don't see why you should be the only ones who have to pick up the pieces."

One night, at the hotel, Mama and I talk about him.

"This guy, Samper, why is everyone fawning on him?"

"Because he's going to be president of the republic, Ingrid!"

"No! That's not possible! You don't mean to tell me that this clown who tells people what they want to hear, who thinks only of making people laugh and yakking with them, is going to be president?"

"Yes, I think he has a very good chance. In any case, the majority of the Liberal Party is already for him."

Eight years later, Ernesto Samper is elected president of the republic, reputedly with the help of money supplied by the drug traffickers. When he is tried by the Colombian legislature for having accepted this money, the situation is clear, at least to me. A person close to one of the alleged murderers testified under oath in another proceeding that she knew Samper had been directly involved in ordering the murders of some who were coming to testify against him. Of course, and no surprise, Samper vigorously denies this accusation. All of these events make me pretty certain about the source of the threats that have been made against me. I am further convinced when, at the same time, *El Tiempo* published an article that revealed a contract between el DAS (the Department of Security Administration) and a private company, Kroll, to spy on and neutralize opponents of the Samper regime.

But in 1986 I'm naïve. For the first time, I'm seeing how our politicians operate, and I understand why my father distrusts them. My mother is less rigid. To my astonishment—and disgust—she replies with resignation that there's a long tradition of corruption in Colombia and that we have to work with these elected officials in order to have an opportunity to be an honest part of the system, if we want to reform it. Ultimately, though I am not yet able to formulate it, I agree with her: Yes, we have to participate. I believe that we can't hand over the country's destiny to men who take no interest in the misery of the Colombian people, who think only about enriching themselves. Yes, we have to change things from within. It is possible. A few days after this trip, I will tell Mama that I'm going into politics, even though I don't fully realize what is involved in this commitment.

The legislature is in full session, and my mother suggests that I come to watch the legislators at work. In the Capitol, I see the coatrooms, the offices, the legislative assembly hall for the first time. I also discover, with great pleasure, how much my mother is beloved by the staff. People open doors for her, embrace her, talk to her about their problems, bring her coffee. At first, I'm seated in a loge reserved for visitors, but before long the doormen, to please my mother, let me sit on the floor of the assembly, alongside her, exactly as if I were a legislator. During my studies in Paris I watched certain debates in the National Assembly, and what strikes me is that here in Colombia legislators speak up all the time, spontaneously, without knowing what they're talking about, purely for the pleasure of getting attention. They seem to me not to have thought about anything personally, to be absolutely ignorant of the stakes. This gradually confirms my impression that most of our representatives are incapable of meeting the expectations of the country.

One afternoon, right in the middle of the session, while my mother and I are sitting there, side by side, I turn to her and say without thinking:

"You know, someday I'll be sitting here myself."

She's surprised; her face lights up.

"Yes," she says, "I'm sure of it."

She squeezes my wrist. We're both moved. How am I going to do it? I haven't the slightest idea. I live in the Seychelles, I'm married to a French diplomat who doesn't want to set foot in Colombia, my whole life resolutely turns its back on the Capitol, and yet I've said this as if I were making a solemn commitment. The words came to me all by themselves.

# CHAPTER FIVE

**THE REALITY** I must face in Colombia during the summer of 1987 is also that my father's heart is giving out.

Late one evening, he calls me.

"Ingrid, tomorrow I have to go to the hospital. Will you go with me?"

I immediately understand that it's something serious; otherwise he would never tell me about it. He's so secretive, so proud.

"Of course, Papa, I'll go with you."

He has to have an emergency operation; his arteries are completely blocked. It's a major procedure and a risky one for a man who's almost seventy.

I don't leave the hospital during the twenty-four hours preceding the operation. We talk; he tells me all his papers are in order, that he's well aware he may be living his last moments on earth. When the nurses come to get him, he utters these joyous but terrible words:

"We'll see each other on the other side of the pass, my dear."

The other side of what "pass"? Death? The bypass operation?

He comes back with tubes sticking out of him, his chest belted, deep in an artificial sleep halfway between life and death. I'm not eager for him to come out of it, because I now see what he's going to have to endure. I stay with him, holding his hand in mine. Finally he slowly opens an eye, closes it, opens it again. He sees me, tries to smile. He wants to tell me something. I bend down and put my ear near his mouth, and I hear:

"Do you know what they found in my heart?"

"No. Tell me."

"Your name."

"Papa!"

I put my hands around his face, I press it against my cheek, I kiss him, and I weep. He's really alive, he has crossed the pass.

Three days later, I find him already sitting up. He's pale and is breathing with difficulty.

"Do you hurt, Papa? What can I do for you?"

"Help me get in bed. I'll be all right."

I take him in my arms. He suddenly becomes very heavy and collapses, pulling me down with him. The electrocardiograph sounds an alert. He's dead! I have the terrible feeling that he has just died. Then I hear myself screaming, as if I were being thrown off a cliff, the constant screaming of someone who's dying, and a second later a dozen people are rushing toward me. They pick up my father, put him back on the bed, and fall violently on him. I seem to be witnessing a strange wrestling match in which the wrestlers are expending enormous energy in order to breathe life back into a figurine, and this gray, inert figurine is my father. They take turns, and my father's heart slowly begins to start up again, a few beats, then the slow rhythm of life that's still there, moving in harmony with the hands of a clock.

They make me go out, but an hour later the doctors tell me I can go back to his bedside. Papa gently takes my hand and says mockingly:

"Gave you quite a scare, didn't I?"

From that moment on, I live with the constant fear that Papa is going to die while I am far away. Ultimately, I'm angry at myself for having left him alone in Bogotá when I decided to rejoin Mama in

Paris, and the idea that he might die in the same solitude is unbearable for me. This idea, always in the back of my mind, will play a major role in the way future events unfold.

Family life begins again in the Seychelles, easy and peaceful, but for me increasingly frustrating, increasingly suffocating. I resume my daily telephone calls to Mama. She's pinning all her hopes on Virgilio Barco, the new president of the republic elected in 1986. She knows him well, for she was his assistant in the social affairs office when he was mayor of Bogotá. He's an intelligent man, extremely well prepared for high office, with great intellectual and moral rigor. He has what it takes to open the Colombian economy to the outside world, to begin peace negotiations with the guerrillas, and at the same time to conduct an all-out war against the drug cartels, whose criminal financial power is beginning to worry people well beyond our own borders, and especially in the United States.

Over the next several months, however, my mother's hopes fade away. Although Virgilio Barco vigorously confronts the drug mafia head on, alerting the international community to the tremendous and ravaging power of the drug lords, going so far as to demand that the law providing for the extradition of drug traffickers be approved by the legislature, toward the end of his four-year term, he seems weakened by health problems. Later it will be revealed that he's suffering from Alzheimer's disease. And Pablo Escobar, the boss of the narcotics traffickers, has already declared war on the government. Despite the Barco government's efforts, Escobar will soon terrorize the whole country. A pitiless bombing campaign strikes blindly in 1987 through 1989 at the cities of Medellín and Bogotá, right downtown, in the supermarkets. Women and children are killed, people are frightened. They can no longer imagine that the state, which is being slowly infected by the mafia, can protect them. Powerless, they watch as their institutions collapse.

Then, curiously, my mother regains her courage. I can tell this from our telephone conversations. Among all the possible successors to Virgilio Barco, one man gives her renewed hope. His name is Luis Carlos Galán. He's not much over forty and, like my mother, is a member of the Liberal Party. Some years ago, he made a name for himself by demanding that Pablo Escobar, who'd succeeded in getting himself elected as a substitute legislator, be expelled from the party and from the Capitol. This morally intransigent man now dares to demand that, right in the middle of the bombing, Colombia sign the extradition treaty for drug traffickers that the United States is requesting. People in the mafia are not worried about going to prison in Colombia because they can quickly bribe their way out. But they do fear being extradited to the United States because they know they might not be coming back.

Galán is aware that by brandishing the sword of extradition he's putting his life at risk, because the mafia is going to sentence him to death, and that is the measure of his courage, his integrity. Mama admires this man who over the months has managed to establish himself as the Liberal Party's best candidate for the presidency of the republic.

When the election campaign begins in early 1989, my mother is once again responsible for the candidate's logistics. A strong friendship grows between them. Mama is older than Luis Carlos Galán, and she feels a protective maternal feeling for him. She often speaks of him to me as if he were her son. She believes in him as she has never believed in any other leader, and she puts all her devotion and energy at his disposal. "Ingrid," she repeatedly says, "he's the last chance for Colombia. He absolutely must be elected."

I continue to experience the Colombian tragedy through my mother. But my country has become a subject of conflict between Fabrice and me. I want to go back, I no longer think of anything else, but Fabrice can't make up his mind to take the necessary steps. But he does accept a posting that brings us closer: Los Angeles. And so we leave the Seychelles for a residence in the United States. Still, the despair at being so far away from my people stays with me.

Lorenzo is born in the United States in 1988, and during the summer of 1989, I go alone with him to France, on the pretext of showing him to Fabrice's parents. In reality, it seems to me that I need to go away in order to think. I'm apparently on the brink of making very serious decisions, but I want to consider them carefully, take my time.

I visit old friends from Avenue Foch days and my studies at Sciences-Po. I take nostalgic trips across France in midsummer, spend peaceful afternoons at the home of one friend or another, dine in the orange rays of the setting sun. On August 18, 1989, I'm visiting friends in the area around the chateaux on the Loire River. I've cuddled a lot with Lorenzo and I'm feeling relaxed and serene. I've never had much difficulty going to sleep; I can doze off almost anywhere, and this evening in particular I feel drowsy. But curiously, sleep doesn't come, and as the hours go by, an inexplicable fear begins to grip my heart. I think about my mother and I'm afraid. It does no good to tell myself that this is idiotic, that since she's been working with Galán she's more flourishing and triumphant than ever. My anxiety persists. I count off the minutes, sitting up in my bed, my stomach tied in knots. Am I having a little bout of depression? I'm going through the strangest night of my life, punctuated by moments of fear that leave me panting and speechless. Then I feel a bodily need to be near Mama, I feel very strongly that only her presence could calm me.

At eight in the morning, when I hear the first familiar sounds in the house and can make a phone call without disturbing anyone, I call Mama. It must be midnight in Colombia. I'm going to wake her, but that's too bad.

"Mama, at last! Excuse me for calling so late, but you know . . ."

And I hear her weeping. For a moment that seems never to end, she can't speak. She's drowned in pain.

"Ingrid! Ingrid! They've killed Galán."

"Oh, no! When, Mama? When?"

"This evening. I was at his bedside. Only about three hours ago."

I am suffering for her, with her. The deep, inconsolable wound of not being able to be near her, to share this tragedy with her and the Colombian people, intensifies my pain. The death of Luis Carlos Galán, for which Colombians are still paying today, is going to mark an irremediable rupture in my life.

Mama explains that she's tried to reach me, too, that she's just talked with Fabrice in Los Angeles, that she was hoping I would call.

Gradually, she regains the strength to tell me what happened.

Galán was assassinated as he was preparing for a public meeting in Soacha, a working-class suburb of Bogotá. That very morning, Mama had tried to get him to cancel the meeting; they'd even argued about it, something they'd never done before.

"I went by to have a look at the place," she tells him. "This meeting is sheer madness. You're going to be in the middle of a public square surrounded by trees, completely open. It's an ideal place to shoot somebody."

She reminds him that the preceding week he just barely escaped an assassination attempt in Medellín. The bomb's timer was incorrectly set, and it went off a few seconds after his car passed by.

"I'm going to Soacha, no matter what," Galán says grimly. "I'm not going to hide out during the whole campaign on the pretext that they're trying to kill me. They're just trying to shut me up, to neutralize me. I'm going down there."

"I'm not telling you to hide, Luis Carlos, I'm telling you that this particular situation is too risky."

"Don't try to talk me out of it. I'm going."

"Luis Carlos, you're refusing to hear what I'm saying. I don't like this place, I'm afraid."

"Yolanda! Until I tell you otherwise, I'm the one who will decide about my meetings. I think this one is especially important, and I'm going, and that's it."

Then my mother, who worships this man, says:

"Don't get upset. I know you'll go anyway. Let's just try to think how to handle this. How are we going to ensure your safety?

Galán calms down.

"I'm worried too, you know. I called the Ministry of the Interior, and they're sending me an extra escort. Ten men. And an armored car. They assured me that the square has been under surveillance since this morning."

"Okay, but you won't be in the car when you're speaking."

"I know, but that's the best we can do. Come on, Yolanda, it'll be all right, let's get to work!"

The meeting was set for eight in the evening. Mama went to the square an hour in advance. She asked her driver to park unobtrusively, and from the car she watched what was going on in the streets. The crowd was already there with banners and steadily growing larger. At all the windows of the houses above, onlookers were hanging out, obviously excited by the spectacle. Mama felt sick; it occurred to her that the place looked like a bullfight arena a few minutes before the bull comes in.

Finally Galán's arrival was announced, and Mama told her driver to park alongside his car. It was in fact armored, but there was no escort. My mother got out of her car, and to her great surprise Galán got out as well. They were immediately surrounded by people.

"You're crazy! Get back in the car! Please, quickly!"

"Yolanda, come on . . . I've planned something else. People have to see me. I'm going to get in the back of that pickup truck over there and go around the square."

"That's complete madness!"

"Stop it! And come with me, the bodyguards are going to get in too."

He was pushed toward the pickup, and my mother, followed by the bodyguards, climbed up behind him, onto the platform in the back. From the balconies, people were throwing flowers to them. Galán was delighted, but my mother saw the danger. They were completely exposed, perfect targets for any *sicario*. Galán moved away from the group and climbed onto a wooden box. He greeted the crowd, opening his arms wide. Mama was beside herself. Then, for just a second, one of the bodyguards took her hand.

"Don't worry," he said. "Feel this."

Mama realized that Galán was wearing a bulletproof vest. That calmed her a bit.

Galán turned around.

"Don't worry, everything is going fine."

He looked happy and triumphant. The truck moved forward, slowly entering the crowd. He waved to the crowd, and people cheered wildly. Galán was charismatic, obviously beloved by the people. Mama stopped trembling, gradually reassured by the delirious enthusiasm that surrounded them.

It was over, and the high-risk parade was a success. His morale boosted, Galán jumped down from the pickup. Now he had to go up on the stage and take the microphone. My mother followed him. She was supposed to sit behind him, along with a handful of other representatives.

While he was climbing the makeshift stairway, his head and then his torso rose above his bodyguards. Mama slipped and fell. At that very moment she heard what she took for a split second to be a firecracker. But when she tried to get up, she was thrown violently to the ground.

"They're shooting at us!" someone cried.

She looked up and saw Galán collapse. His personal bodyguard, already hit by several bullets, uselessly tried to cover the lower part of Galán's body.

They took Galán away. Surrounded by the escort, my mother was pushed behind the walls of the nearby town hall. The attack was immediately announced on the radio, and soon it was said that Galán was not dead, but that they would certainly need O-negative blood for a transfusion. This is Mama's blood type. An ambulance, which was still there, took him away. Mama and the other members of Galán's staff assumed that he would be taken to the hospital closest to Soacha, but they didn't find him there. They asked where he was, and finally, their sirens screaming, they arrived at the right hospital.

There, the panic and confusion were total, with cars parked every which way. Galán had just been taken out of the ambulance and was still lying on a stretcher. Mama heard a frightened nurse cry:

"What happened? Who is this man?"

"It's Galán! It's Luis Carlos Galán for God's sake! Quick, I beg you, he's dying."

They ran and succeeded in giving him a transfusion. Mama was the only one of his close associates at his bedside; his family had not yet been able to get there in time. He died a few moments later without having regained consciousness. It was all over.

My mother had just come back from the hospital when I called her from France. For her, Galán's death was the death of Colombia. She was sick with pain, in deep despair. He was the last obstacle capable of resisting the disease of corruption that was infecting the whole apparatus of government. We talked for two hours, and I heard myself repeating, despite myself, an unbearable recognition of my culpability: "I should've been there, Mama. I should've been there with you."

---

Four months later, I make my decision. I leave Fabrice, pack my bags, and fly alone to Bogotá. I'm well aware of the suffering that awaits me, the separation from my children, the pain of not having been able to save my family—just as Mama, by an irony of fate, had failed to save hers fifteen years earlier—but I'm sure this is the price I have to pay to finally recover a place among my own people.

# CHAPTER SIX

**WHEN I ARRIVE** in Bogotá in January 1990, my mother has just decided to seek election to the Senate. She's heartbroken, deeply discouraged, but she's doing this to honor the memory of Luis Carlos Galán. Since I've made a clean break with my earlier life, for me her commitment represents, even in the horrifying context of Colombia, a promise for the future. It's a modest but blazing torch I would like to help her carry if I am strong enough.

I'm twenty-nine and I have neither job nor money, so I move in with her. At this time, my mother is the only link binding me to the native country I left more than ten years earlier. I've lost contact with all my friends from the French school, and I don't know any of the current political stars personally. Curiously, I find this isolation stimulating. There is so much for me to do, to learn, and the desire is there, amplified by a decade of waiting and dissatisfaction. I feel strong, self-confident, and ready to fight. Especially for my children. To have them with me. Naturally, I miss Melanie and Lorenzo; some evenings are terrible. But soon Fabrice sends the children to spend a couple of months with me in Bogotá and then I feel complete.

Fabrice and I will not escape the painful, turbulent conflicts of most couples who separate, but he acts very honorably, and about a year later he comes to live in Bogotá, for the sake of the children.

What joy it is for me to finally be in Colombia! Mama talks freely. With me, she no longer has to pretend, she can open up, she has nothing to hide. From her description of the situation, I can see how bad things look for the country. The political leaders whose faces I see every morning in the newspapers seem lacking in stature, without ideals, interested only in power and money. Now we're approaching the crucial dates on the election calendar: after the reelection of the legislature in March comes the presidential election, also held in the spring. Galán is gone, and people are looking for possible replacements. Samper's name has been mentioned—Ernesto Samper, that caustic and truculent man who made us laugh so much during the trip along the Atlantic coast four years earlier. But the Liberal Party finally chooses Cesar Gaviria, who's in a better position to take advantage of Galán's political legacy because he was his campaign director.

Mama finds herself caught in a terrible dilemma: she has no confidence in Gaviria, whom she suspects him of being too "flexible" with regard to principles; on the other hand, she has no choice but to support him. He's the candidate of her party, the Galán family has officially endorsed him, and no providential man has stepped forward. "As often happens in Colombia," she will say, "we're reduced to supporting the candidate who is least bad."

Soon my mother throws herself into these two successive campaigns: the conquest of her senatorial seat, and then, for Gaviria, the presidency of the republic. And I become more than her escort and confidant; I gradually become her advisor. Everything I've learned at Sciences-Po, everything that has fed my passion for governmental administration, comes rushing back to me. Together we design posters, think about her speeches, the themes she should develop, the words she should use to persuade people. In Colombia, a senator has greater authority than a representative, mainly because it takes many more votes to be elected to the Senate than to the House of Representatives. The election is difficult, but Mama finally wins, thanks to the votes of the most disadvantaged people.

Cesar Gaviria also wins, two months later. But even before he's elected, he has shown another face, in what my mother considers to be a betrayal of Galán, by reneging on the commitment he'd made to the Colombian people to sign the extradition treaty for drug traffickers. Though disillusioned, Mama finds the strength to raise a storm within the party's directors meeting. She succeeds only in making bitter enemies, and soon she feels quite lonely following Galán's ethical line.

As for myself, when the elections are over I'm out of work again. But these weeks of meetings, traveling, and rushing about have allowed me to reconnect with old friends. For example, I've become reacquainted with an old school chum, Mauricio Vargas. Just as brilliant as ever (he was always first in his class), Mauricio was already, despite his youth, the director of one of the largest Colombian weeklies, *Semana*. As soon as Gaviria's government is formed, Mauricio calls me to say that he's recommended me to the new minister of finance, a friend of his, and that the minister has agreed to see me. My choice is made. I don't want to go into business; making a lot of money doesn't interest me. I want to take part in running the country. I keep thinking about what my father said while we were on the ocean liner coming back from France: "Ingrid, because you've had so many advantages, you now have a debt to Colombia. Don't forget that."

The minister of finance is named Rudolf Hommes. He doesn't come from the incompetent and corrupt Colombian political oligarchy—and this is my good luck. He's a professor at the university, known and respected in financial circles—a technocrat. His reputation recalls my father's when he was a minister: people say he's competent and rigorous. Coincidentally, the ministry of finance is now situated in the building the ministry of education once occupied, and Hommes receives me in my father's old office.

We get along from the start. He's a small, stout man with bright blue eyes. He asks me to tell him about my studies in Europe. He finds my education appealing and rather romantic. Very quickly, I sense he's surrounded by a team of brilliant young professionals, all of them trained in the United States. I'll be the only member of the team who was trained in France.

"Okay, then—you haven't forgotten how to add and subtract?" he finally asks me.

"I think I can manage."

"Fine, you're hired. You begin tomorrow."

"What precisely am I supposed to do?"

"You'll see."

The next morning I go into my office. It's small, but right next to the minister's office. I'm one of his technical advisors, and I have a secretary—a precious secretary, well acquainted with how the Colombian government works. I've scarcely sat down before the minister calls me into his office.

"Get in contact with the NPD regarding this file and give me a report."

The NPD? I don't even know what it is. My secretary bursts out laughing.

"It's the National Planning Department, Ingrid. I'll call them for you."

If I have had too high an opinion of myself, I suddenly understand how my ten years out of the country have handicapped me: Not only do I lack contacts, but I don't know the codes; I lack all the reflexes familiar to the most humble Colombian student. I'm a foreigner in my own country.

Over the days that follow, the incessant ballet of visitors prance around the minister. Ministers and ex-ministers, ambassadors, politi cians, officials from international organizations, and other well-known people all stop by my office to say hello.

"So you're Yolanda's daughter! Be a dear and tell her that I'm thinking about her and send her kisses."

Or:

"My Lord, you're Gabriel's daughter! I admired your father very much. Tell him how sorry I am that we never see each other any more."

All these friendly, charming people are apparently convinced that I know who they are, whereas I haven't the slightest idea what their names are or what high offices they occupy. I feel as if I've just emerged from a long bout of amnesia, and sometimes the void in my memory makes me dizzy. If they only knew, I say to myself, how lost I feel, how ignorant I am of the bonds that form their membership in Colombian society, they'd turn away from me with embarrassment or discomfort, the way one turns away from an intruder in a family reunion.

Nonetheless, I like working, and I try to convince myself that in this job, thought and method are more important than public relations. In particular, I'm convinced that the only way to make progress is to propose a solution for each problem. Never to shelve or slow down an issue on the pretext that it presents what seem to be insurmountable difficulties, but to do research, move heaven and earth if necessary, to solve the riddle, undo the knot. That's what I'm stubbornly doing in my little corner, and my minister is aware of my effort to construct viable solutions that often require unexpected collaboration despite the sluggishness of administrative processes.

Between my minister and me a friendly but cautious relationship develops. Hommes feels I'm too independent and likely to take off in a direction he wouldn't necessarily endorse. Is it to test me that he ends up giving me an assignment that no one else wanted?

"Ingrid," he says to me one morning, "I've just told the legislature that I'll propose a plan for the development of the Pacific coast.

It's a complicated issue. Take a couple of months, develop a policy, and give me some proposals with figures."

The Pacific coast of Colombia stretches from the Panamanian border to Tumaco, near the border with Ecuador. I've never been there, and neither has my minister. In fact, very few people are acquainted with this area, for the good reason that no real road goes there. If the region is arousing a great deal of interest, it's because it may benefit from the future development of maritime exchange with the west coast of the United States, Japan, and China. Our Latin American neighbors in Chile, Peru, and Ecuador have done this with their coastal land. As it is, our Pacific coast is a tropical forest of inestimable ecological value; it is the lungs of Colombia. Should we destroy part of this forest in the name of economic interests? Or should we resist the pressures of our businessmen and greedy politicians in the name of protecting our ecosystem?

I soon discover that in the National Planning Department, the NPD, two young technocrats are already working on a development plan. We decide to work together. Before long, this becomes known to the government and political circles interested in the Pacific Plan and the pressure on our little team starts to grow. One man, representing a major project for exporting shrimp, offers to take us to see the site; another also offers to serve as our guide, because he plans to make a fortune by selling pineapples or bananas. The elected representatives of Cali and Medellín, two large cities dreaming of vast commercial outlets in the Pacific, compete for our attention. New invitations come in every day.

We can see danger looming, and one fine morning, fleeing these much too attentive people, all three of us throw on our backpacks and set out for Buenaventura. An old patriarch who lives in the area and whose sole ambition is to help his community before he dies serves as our guide.

The road from Cali to Buenaventura is impassable for the few days ahead (although it is the only access to the Pacific coast), so we have to take a small plane. Our guide is waiting for us. How does he expect to transport us during the ten days we plan to spend there? All he has is a flat-bottomed boat, of the kind that is rented in the Bois de Boulogne in Paris, but equipped with an outboard motor. But the weather is unfavorable, as is often the case on the Pacific coast: dark, rough seas, and a leaden sky.

We set out. The boat takes in water and there are no life jackets. At that moment, I wouldn't bet my shirt on our chances of surviving. However, a minute later we leave the high seas and head for the mangroves through a maze of little waterways. We enter the mangrove, which we will not leave again. Here the water is calm, not very deep, and full of an intense, secret life whose echoes we faintly perceive: crabs, langoustines, fish, and animals of the tropical forest live there side by side. The sound of our little outboard motor causes colonies of birds to rise into the air, and we have to stop the motor to hear the continual rustling in this moist, exuberant vegetation. Fine, but where are the people? How do they manage to survive in these lukewarm swamps, so far away from everything?

After moving on for another three hours, we spot a handful of cabins against a background of leaves and creeping vines. They're just shacks, and they seem to be uninhabited. Our boat, however, turns toward a strip of black, muddy earth. A dozen children appear. They throw down a tree trunk to serve as a bridge to this strip of land, and we climb out. Two men observe the scene from a distance.

The men are very tall, dressed only in shorts. As we start to walk toward them, buoyed by the children's euphoria, the door of one of the cabins opens and a woman comes out. In these surroundings, she seems completely unreal: an astonishingly beautiful green-eyed brunette, her body encased in an immaculate T-shirt, pants, and high heels.

"Our teacher," the children say.

Clearly, we're expected. The young woman smiles at us, invites us in. Makeshift desks, maps. Look here, Iraq and Iran. It's January 1991, and there's a war over there. Could this teacher, here in the depths of the tropical forest, possibly . . . ?

Yes, in spite of the conditions and the distance, she says, the children have to understand the world they are living in. We talk, and I learn how this woman came here from Cali, where she has left her husband and her children for the time being. There are only two ways to get a teaching post in the city itself: either she'd have to pay part of her salary to the local politician who got her the job, or she'd have to go to bed with him. Since she refused to do either, she was punished by being sent for a year to this muddy hole. Instead of giving up, she decided to meet this impossible challenge: teach the children and bring a minimum of culture to a community that hadn't had a teacher for a very long time.

"When I got here," she tells me, "the children were drinking water from the river and many of them were dying of diarrhea. I convinced the families to keep a pot of water boiling all the time." She set the example, kept her own fire going. Then she had to get them to dig sewer canals, to protect themselves from mosquitoes, and to respect privacy—for at first both the children and the adults rushed to see the teacher bathing in the river. The Colombian government has done nothing for this village other than send it this teacher, and she, who by rights should despise the government, is doing its work for it. I know that someday Colombia will get on its feet, thanks to the silent effort of all these men and women, honest and strong, whom I've met throughout my country. Their eyes will always be on me, like an oath that has to be honored.

We arrive at Lopez de Micay, a village whose church steeple towers over roofs and mangroves. Ah, here at least the government has built a dispensary—the first such achievement we've seen. Two young men who are completing their medical studies have been assigned to it, but they have nothing to work with, not a single instrument, no medicine, not even a package of cotton.

"Do you want to see?" asks one of them, speaking in a slightly scornful tone because he assumes we're the usual pencil pushers.

"Yes, please."

He holds up an aluminum box, then opens it: it contains a syringe and a rusty needle.

"That's it. That's all."

Then, by an instructive and dramatic coincidence, a family rushes into the room where we're talking. The woman is about to give birth, but she has to have a Caesarean delivery. The dispensary has no boat to transport her to Buenaventura. We lend our own, in the hope that in Buenaventura a doctor will have the necessary equipment to save this woman and her children. What kind of a democracy is it that lets its people die like this without any choice?

The dispensary itself has not been finished. Why? Because the local representative who obtained the funding for it skimmed off half the money for himself and his pals. He'll be able to boast that he got a dispensary, even though this dispensary is useless. By the same cruel irony, he'll also boast of having got a dental office, whose high-tech chair has in fact been delivered, but not the electricity necessary to use it, because there is no power plant at Lopez de Micay yet.

This is a double misfortune for the people: not only does the government fail to help them, but they have the feeling that it manipulates them as well, that it's toying with their innocence and wasting the money that would be so precious to them.

Once our boat is back, we continue our journey. A short time later, we arrive in a hamlet, El Charco, that is in the midst of a catastrophe. Most of the houses have just burned down, and people are running around in a suffocating cloud of smoke. They finally throw pails of water on the fire. The town is relatively large, with a real landing stage and a dirt street that has actually been graded. In the panic, no one pays any attention to us, and when calm has returned we see children lining up in front of one of the buildings still left standing: the grocery store. The grocer is handing out free food, as the government would in other countries. Later, families are preparing dinner. Though they've lost their homes, they don't seem to be in complete despair: they eat, they talk to each other quietly, the children laugh. Having come from the ministry of finance, we're almost ashamed to arrive empty-handed. But they're friendly and warm. They tell us they're going to rebuild all the houses that have burned down. They seem to expect nothing from us, and when we succeed in convincing them that we've come to tell them that the Colombian government owes them at least minimal assistance, the only thing they ask for is a little boat with a motor that would allow them to reach the hospital in Buenaventura quickly. Children, pregnant women, and elderly people are dying because they lack such a boat.

Finally, we arrive in Tumaco, our last stop. And this time we see the money, palpable, obscene: luxury cars, private yachts, flashy houses, like palaces from the Arabian Nights, standing along streets that are not even paved. In Tumaco, large business enterprises are involved in exporting and importing, and unbridled smuggling is also carried on, especially by the local elected officials. Alongside this ostentatiously displayed wealth, a poor neighborhood is built on stilts over the water. That's where the workers and their families live, the ones who work in these profitable businesses. Three thousand people crammed into shacks eaten away by the damp, over a heap of garbage rolled endlessly back and forth by the waves.

Promises to provide new housing for them are constantly made, but nothing is done. The senators and the representatives, members of the same clan who get a commission on everything that happens in Tumaco, are in no hurry to bring these people back onto terra firma. We meet the priest, who's leading a lonely battle on behalf of these abandoned people, and then we learn the extent of the criminal negligence on the part of the representatives of Tumaco: tidal waves are frequent, and these fragile houses on stilts can't stand up to them. At least once every ten years, the horror is repeated: the shacks disappear, carrying along with them the men, women, and children who live in them. The survivors and newcomers, driven by the same poverty, rebuild on the same site, no doubt preferring the ferocity of nature to that of their fellow men.

We have to do something. When we get back to Bogotá, we write it all down, inform the government, knock on every door—all in vain. No one does a thing. No one cares. Years after this trip, after I've been elected to the legislature, a tidal wave destroys this whole neighborhood in Tumaco, killing hundreds of people. That day, as I fight back my tears, I think that the most urgent of all battles, and the most legitimate, is to give Colombia true leaders worthy of the name.

# CHAPTER SEVEN

HAVING BECOME AWARE of the crookedness of the officials, of the systematic diversion of funds, and also wanting to preserve local identity, we draw up an essentially ecological development plan that tries to allow local populations to make the decisions that affect them. Rather than pursuing pharaonic investments sought by the politicians and wheeler-dealers, we argue for communitarian development projects using local materials, especially water supply and sewer projects. We emphasize the effort on schools and health care systems. Soon, as this program's sole proponent, I'm traveling back and forth between Bogotá and Cali. My minister supports me, and while I fail to persuade certain officials who are too closely connected with the local elected representatives, I win the support of a significant portion of the press and especially of the target population.

One day, the governor of the Cali region invites me to the opening ceremonies for a new low-cost housing development. There I meet with the minister of development, who is no other than Ernesto Samper.

"Oh, Ingrid! What a nice surprise! How are you? You're magnificent, more and more like your mother."

The same old Samper, warm and charming.

"What are you doing here?" he asks.

When I explain that I've completed a development plan for the Pacific coast, he suddenly becomes very attentive to what I'm saying. For a man usually so jovial and flighty, this is a rather good surprise.

"Will you give me a copy?" he interrupts.

"There's been one on your desk for almost two months! You can easily imagine that as minister of development, you were one of the first people we sent it to."

"Ah, good, good, I'll read it as soon as I get back. But go on. You were saying that so far as low-cost housing was concerned . . ."

I explain every proposal and its projected budget. Then we separate. He has to give a speech that afternoon and I have to attend a series of meetings elsewhere.

How surprised I am the next morning when I open the newspaper! On the front page, I find this headline: "Minister Samper Launches Pacific Plan: Millions of Dollars in Investment." Everything I had told him the day before is there, crudely summarized. Naturally, my first reflex is to laugh at the guy's incredible gall. From a single fifteen-minute conversation, without the slightest reflection, avoiding any previous government consultation, he has derived this enormous announcement that concerns tens of thousands of people. But my laughter soon turns to exasperation. Samper has shown that he's insufferably cynical; usurping the project of the minister of finances, he flouts a government that tolerates him only in order to avoid a split within the Liberal Party. In the absence of any personal convictions, his strategy for taking power is based entirely on bluff.

For Rudolf Hommes, my minister, who's been supporting this project from the outset, this is a slap in the face, and as soon as I return to Bogotá I get an unprecedented dressing-down. But the worst thing is not this tirade, which we will both soon forget; it's that Ernesto Samper, by hastily making this announcement, has killed the Pacific plan before it ever got off the ground. Why? Simply because the head of state, Cesar Gaviria, who has detested Samper ever since they were competing for the presidency of the republic, does not want his minister and rival to get an electoral advantage from this program for development. So Gaviria will do his best to minimize and then bury all these measures the local population had begun to hope for.

Then I'm given a second assignment as thorny as a cactus: to address the smuggling issue. It's urgent: Colombian industry is dying because of the fraudulent entry of foreign goods that are not taxed and thus are sold on the market at prices significantly lower than our own local products. This is especially the case with the Colombian tobacco industry, which is being choked off by contraband American cigarettes. Our textile, shoe, and alcohol products are also being threatened. On the other hand, a considerable number of people live off this illicit trade and it would be irresponsible to try to change things without providing any legal alternative to them. So our idea is to delineate geographically the so-called free trade zones within which Colombian products would be sold free of tax in order to neutralize the attraction of contraband. Why should anyone take the trouble to import American cigarettes fraudulently if our own are sold for the same price or even less? This transitional policy would allow the local population to legalize trade while at the same time putting an end to the criminality inherent in smuggling and in the network of corruption it requires.

However, our plan will go into effect only if enough people can be rallied to support it. And so all three of us, three technocrats, set out for Maicao, the contraband capital of the Atlantic coast, the very place to which I had accompanied my mother on that surreal journey with the legislators in the summer of 1986. This time, the mood is no longer festive. A Toyota panel truck is waiting for us when we get off the plane, and it's riddled with bullet holes.

"What happened to this car?" we ask.

"It's very dangerous here," the driver grimly replies. "You have to be careful, people are violent."

Just in case we haven't gotten the message, a giant banner hanging over the road into Maicao repeats it: "Ministerial envoys get out!" To complete our welcome, all the businessmen have pulled down their metal blinds. Operation dead city, in other words, and everywhere we see posters: "Get out!"

A public meeting is set for 5:00 P.M. in the Commerce Club, an enormous shack with a floor of beaten earth that can hold up to five hundred people. When we arrive, the atmosphere is highly charged. A crowd of men is looking at us suspiciously—I see only one woman, wearing the *wayuu* costume of the local Indians—and, more troubling, bottles of whiskey are being passed around. The air is heavy, and we're all sweating.

Leonardo, the only man in our trio, begins by explaining the project from a technical point of view. He's greeted by a sardonic silence. Then hands shoot up and people begin to speak. From six o'clock on, they almost monopolize the microphone for more than four hours. Leonardo manages to get only a few words in. Gradually, the tone grows more hostile, stimulated by alcohol, anger, and the heat, until it culminates in expressions of hate and insults: "The government is laying a trap for us." "All this just means more taxes." "You are liars." "You are not welcome here."

When Leonardo tries to regain control, men who are completely drunk spew saliva, shake their fists, and come up and threaten us: "Liars! Little bureaucrats! You don't know anything about this area, get out of here!"

"Let's leave," my colleague whispers in my ear. "We're going to get ourselves lynched."

In my view, running away now would be a disaster for the image of the government, which is already not so good. So I decide to play the sole card left to us—an appeal to gallantry, which always touches the hearts of Colombians. I take the microphone and stand up.

"I see only men here," I say solemnly. "We've been together for hours, and during all this time I've been waiting in vain for one of you to have the courtesy to at least allow us to say something. But no one has shown that basic politeness."

Silence. It's as if these words had sobered them up. Grumbling, they settle down, and then grow quiet. All I have to do is continue.

"What do you want to defend? The right to kill each other in order to do business? We were brought here in a car full of bullet holes. Is this the climate in which you're forcing your wives to live and bring up your children? What's your idea of the family, of happiness, of life? The richest among you barricade themselves behind barbed-wire hedges, fences, and surveillance cameras, and even they are never sure they're going to make it home to sleep in their beds. What good does it do you to be rich if you have to live like this? Look at your town; it is a mess. No paved streets, no water supply, no sewage system. The truth is that you're all prisoners of violence, of corruption, of contraband. Ask yourself for a minute if your wives and children wouldn't prefer to live with a normal, honorable businessman, one whose success doesn't make him a target. I want to tell you one thing: we haven't come here to sell you something at any price. This free trade zone is in the interest to you all. But we won't enact it against your will. Never! If you want it, fine. If you don't want it, you'll remain in this lawless state in which hundreds of you are killed every year. Personally, in all this I'm concerned mainly about the welfare of those who cannot choose: your own kids."

They look at each other and calm down. They're less distrustful. It's already late, however, so we decide to meet again the next day, not with five hundred people, but with a delegation chosen among them.

Six months later, the free trade zones of Maicao, Urabá, and Tumaco are created by decree.


I'm summoned by the young minister of foreign trade, Juan Manuel Santos. His ministry has just been created in order to help open up the country commercially, and Santos symbolizes tomorrow's generation of politicians: he has a degree from Harvard, he's enthusiastic about globalization, he knows all about advanced economies. The heir of a great Colombian family, he has given up the top position at *El Tiempo*, the main national daily paper, in order to head up this ministry. He's aware that Colombia can't go on as it is, turned inward on itself, sheltered by its borders. To get on the trade bandwagon, it must immediately adopt international regulations, in particular the regulation that respects industrial property—that is, patents.

Santos has called me in to get this legislation passed. Here I discover one of the main reasons for Colombia's backwardness: because we've consistently refused to sign anything regarding respect for patents, international industries, researchers, and innovators have turned their backs on us. Worse yet, we have a reputation as looters and pirates. If a pharmaceutical product appears on the world market, instead of importing it legally, we copy it. I visit sordid garages in which, without any supervision or sanitation, chemists make medicines that they claim are identical to those produced in the laboratories of Europe and North America, although it's likely that they aggravate problems and perhaps even kill more often than they cure. We have a "do-it-yourself" mentality. We think we're being clever, but we don't understand that our lack of respect and own homemade copies, built into the system, have deprived us of all the fruits of the latest industry research. The result is that products "Made in Colombia" are not up-to-date compared with what is being made elsewhere, and thus are not exportable.

Subsequently, I undertake a tour of the Pacific Rim countries—Japan, Hong Kong, Korea, Taiwan—in order to learn how they've developed their international trade and also to present an image of a new Colombia: a Colombia eager to play by the rules and to abide by common laws and ethics. I'm supported in my efforts by our best industrialists, notably the flower-growers, who urgently need this new legislation in order to import new breeds, especially from the United States and Holland, as well as to break into foreign markets.

More than two exciting years go by in this way. Then one day Juan Manuel Santos asks me to accompany him to the ministerial council, held first thing in the morning. One of my colleagues and friends, Clara, goes with us. Santos is supposed to make a report, and we are there to provide him discreet support. It is there that I find out how the council functions. At first I'm dazzled by President Gaviria's intelligence, sharpness, and excellent knowledge of the issues. For each problem, he does a perfect job of outlining what's at stake, showing his precise understanding of the mechanisms of the world economy as well as his extensive general culture. Yet when the time comes to make a decision, he suddenly interrupts an ambitious argument in order to accept a political compromise that seems to me mediocre. Clara shares my impression. We have the feeling that Gaviria and his ministers have a clear vision of what should be done, but that they remain subject to unrevealed pressures and secret allegiances that in every case lead them to draw back and resort to solutions incompatible with the modernization we are dreaming about.

Clara and I leave the presidential palace deeply disappointed, and before returning to the ministry we go to a café to have a real breakfast. As we review the debates, it seems obvious to us that the country's interests are constantly being made secondary to satisfying the interests of people whose influence behind the scenes we clearly sense.

"It's terrible," Clara says, "everything we propose will always be blocked by the same lobbies that have an interest in keeping anything from changing."

"Unless we decide to move very fast."

"What do you mean, 'very fast'?"

"I don't know . . . We're technocrats, Clara. We have the right, and even the duty, to suggest solutions. But we don't have the power to implement them. We don't have any power, really. The true power is in the hands of the politicians."

At that point, we part. But a week later we meet for lunch and pick up the conversation where we left off.

"I've given this a lot of thought, and I don't see why we should continue to work under these conditions."

"Are you thinking about moving over into the private sector? We'd make a lot more money, that's for sure," she says laughingly.

"Come on, you know we're not doing this for the money! No . . . what if we went into it ourselves, Clara?"

"Went into what? What are you thinking about?"

"Going into politics, of course!"

"I've thought about it, but it's against the rules. You can't hold a government job and be a candidate for office at the same time."

"Are you sure?"

"More or less, yes."

"Let's find out. Okay?"

It's August 1993, and the legislative elections are scheduled for March of the following year.

That afternoon, we send a long letter to the electoral commission. Clara is the director of a department in the ministry, and I'm a technical advisor to the minister. Do we have the right, holding such positions, to present ourselves as candidates for seats in the House of Representatives, for instance?

A month later, we receive a reply: yes, we are authorized to present ourselves as candidates—but on one condition, which is not negligible for people like us who have no personal financial resources: first, we have to give up our jobs at the ministry by next Christmas, that is, before the end of the year.

# CHAPTER EIGHT

"GO AHEAD, INGRID! This is the time. You have experience in government and you're at an age when you can be the embodiment of renewal. I've always thought you were made for a political career. And I won't get in your way, either, I'll withdraw. After Galán's death, I no longer have the energy and the faith to continue. Go ahead! It's your turn."

"All right, Mama. But how do people go into politics? I mean, in the real world, how do you do it?"

"Listen, my own case was special. The public knew me before I went into politics. I don't know how other people do it. But go see José Blackburn and tell him I sent you. He was on Galán's team. He's still a friend of mine and he knows all about it."

José Blackburn was at that time a powerful constructionist. He was an active Galán supporter and got himself elected to the legislature, first as a representative, then as a senator under the banner of the New Liberalism. He listens to me sympathetically and tries, as if he were my advisor, to direct me toward a more realistic goal.

"It's hard to get elected as a representative, Ingrid. Why not begin by getting on the Bogotá city council?"

"No, I want to work for the country, to change things at the top. Otherwise nothing will change."

"You're very ambitious. And very idealistic as well. Well, why not, you'll be a young, new face in a cast that remains largely the same. And then you're a woman, and that's an advantage. Listen, the first thing to do is to get together enough money to finance your campaign. The second is to find a place to set up your headquarters. The third is to round up some people, as many as you can. Take your address book and call every person in it, even people you've met only once, and convince them to come help you. You need people, Ingrid. You can't do anything alone. Or else you can buy votes, like everyone else, but to do that you need lots of money, too."

"You're out of your mind! It's precisely to put an end to that kind of thing that I'm going into politics. I want to prove that it's possible to be in politics without having to buy people!"

"Good, very good! There, you've already got a program!"

That evening, I see Clara again. She and I have solemnly agreed to go into politics together or not at all. For the moment, we're standing hand in hand at the edge of the abyss. It's delightfully dizzying, but we haven't done anything irreversible so far. We can still retreat to our ministerial firesides.

"Money?" Clara says. "Well, I'll tell you what: we're going to set up a work session, the kind we've learned to organize here. We'll invite the industrialists who know us from the ministry, and we'll put our cards on the table. They're sure to have ideas, and anyway, it will provide a good test."

So we organize our first breakfast meeting as potential candidates, and invite the ten CEOs who've been closest to us during the last months, especially the flower-growers. I know they admire our professionalism, and we were already linked by mutual respect.

"We've been working in the wings of government for three years now," I tell them. "On each of my assignments, I've proposed solutions directed solely toward the interests of the country. However, with few exceptions my proposals have been cut back, diverted, or simply set aside, by the very people we've elected to make such reforms: the politicians! Today the Colombian people feel powerless when confronted by these corrupt elected officials. They claim to be taking our destiny in their hands when in fact they are stealing it from us. I want to show Colombians that it doesn't have to be that way, and that politics can be practiced differently, especially by getting elected honestly, as in mature democracies, on the basis of your ideas, with posters, speeches, and a program—that is, without buying anyone and without putting yourself up for sale. In short, I plan to resign from the ministry in order to become a candidate for a seat in the House of Representatives, and I need money. But I want to make this absolutely clear: I will give nothing in exchange for financial support. I will simply work to construct a democracy worthy of the name. And it will take a long time."

A great silence. Is this an illusion, a laughable illusion, to think that at thirty-two I can reform, all by myself, a system that has been in place for decades? Could these men, who are for the most part twenty or thirty years older than I, agree to help me, to believe in me? Some of them have amused, perhaps incredulous smiles on their faces, but no one has gotten up to leave. On the contrary, they seem to be looking at each other as if they were conferring, and the first questions are asked.

These questions are very touching, because, beyond any skepticism, we can hear in them an almost irrational desire to believe, a mad desire for this to work, for us to succeed, despite our naïveté and the inadequacy of our means. These men, experienced in business and in economic warfare, are trying to warn us about the obvious hazards and beginners' mistakes, to get us to focus on the main point right away: how to convince Colombians to vote for us. Time is running out, especially because we're unknown and the election is around the corner, and that's the first thing that worries them. They ask tough questions that force us to formulate our ideas more effectively, so that our breakfast meeting takes on the appearance of a major oral exam. But by noon our program has been formulated and summed up in a ringing profession of faith denouncing corruption. And these ten men, acting really as midwives, without making any demands, will allow us to give birth to our campaign. It's an unprecedented, magical moment; they open the doors of the world to us and then solemnly go away, just as they came. Everything remains to be done, but we're no longer alone.

Now all I have to do is resign my post at the ministry. I'll do it in November, after winding up a series of trade negotiations with our trade partners. I expect my minister to encourage me. After all, he also has political ambitions—he sees himself as president of the republic within ten years—and so he should be able to imagine what's going through my head. But he throws cold water on me.

"You're completely nuts, Ingrid! What you're saying is utter nonsense. Who knows you? You won't get a single vote. Listen, I'm sure your concierge doesn't even know your name. It's grotesque! And I need you here. Think about this calmly and we'll talk about it later."

"I've already thought about it. I'm leaving."

He shakes his head and looks sincerely distressed.

"Well, if you want go to the front, it's your funeral. All the same, I'm going to do something for you: after the elections, I'll give you back your job at the ministry. That way at least you won't be unemployed.

"After the elections, I'll be a representative."

"Sure, sure. We'll talk about it later."

But I'm not so certain, though I've made up my mind. I have only four months to try to win a seat in the House, and these four months include the holiday season at the end of the year, which in Colombia is the equivalent of the summer vacation in Europe and the United States. That is, we won't really begin our campaign until the middle of January. Then all I'll have left is eight short weeks.

# CHAPTER NINE

SO FAR, so good, but now we need to set up a headquarters for the campaign. We have to hurry, in fact, because it has to be ready when the vacation ends in January. Clara and I drive up Septima Avenue, one of Bogotá's main arteries, searching for an office to rent. Finally one afternoon, we come upon one of the abandoned, ostentatious palaces modeled on those of nineteenth-century Louisiana. This private residence, with an imposing, columned entrance, somehow survived the wave of demolition in the 1980s and is now empty. We park the car.

"That house?" a worker responds to our question. "It belongs to the notary who works on the street right behind it, there."

Well, then, let's go see the notary! We're dreaming? Yes, but why not? The notary's office is very old, dark, and full of people.

"We'd like to see the notary." We find out it is woman.

"May I have your name, please?"

"Ingrid Betancourt. We need only a few minutes of her time."

After a moment, we're ushered into her office: piles of yellowed papers and, dimly lit at the back of the room, the equally yellowed face of an elderly woman bent over her files. She lets me talk, and then says coldly:

"That house is not for rent, madam."

"I understand, we've been told that you're planning to move your office there. But maybe we could pay part of the cost of renovating it."

"There's no point in insisting. I'm sorry."

"Listen, for us this would be such a boon! We're beginning a campaign for election to the legislature. Wouldn't you like to help us a little bit? We need it for only three or four months."

She looks at me carefully. Then we hear her laugh.

"You wouldn't be a Betancourt, would you? What's your father's name?"

"Gabriel."

"I thought so, you're Gabriel's daughter. Well, you know what? I was also in politics, with Dolly."

"Papa's sister?"

"Precisely. Your aunt Dolly. Well, look here."

She shows me a photograph in which I recognize Dolly, and, to her right, the notary, slim and young at the time. They're both wearing the Conservative Party's blue uniform. Dolly, who was elected to the Senate in the 1960s and died in 1974, when I was thirteen, used to hug me when I was a child. People would say that I looked a lot like her. I remember her sweet, distant smile.

"Why on earth are you getting involved in such a mess?" she asks me. "It's hard, you know. Especially for a woman."

I explain. I say that we want clean politics, and we're not going to quit, we'll go all the way against corruption, against the mafia's takeover of our institutions, of democracy, all the way.

She listens. She smiles, almost tenderly.

"Well, I'm going to do something for you, and for the country as well. In the name of solidarity among women. You've got this house. Go ahead! We'll talk about money later."

After we leave her office that day, standing on the sidewalk, we feel our hearts exploding, we feel like screaming with laughter and crying.

If we've got this house we can't fail to be elected! Absolutely can't! This magnificent lady believes in us to the point of giving us this gift. Ten important industrialists trust us to the point of writing a check for us. Doesn't this mean heaven is on our side? Clara and I hug each other and astonished drivers in passing cars honk their horns.

An hour later, we're feverishly going over our new palace. Two living rooms on the ground floor capable of holding at least a hundred people apiece for our public meetings, and about fifteen offices on the upper floors. Clara and I will share one office, otherwise we'd lose the habit of talking to each other. Here, we'll set up our future press attaché, here the accountant, here the telephone switchboard . . . Yes, but in the meantime the rooms are empty! We'll need hundreds of chairs, cabinets, lamps, desks. We can't possibly buy all that, it would virtually exhaust our budget. Quick, the telephone book. Look, there's a furniture factory nearby—let's go right away.

The man has the austere face of an old carpenter who's seen everything. Our excitement troubles him not in the least.

"Give me your address, I'll come by tomorrow."

He's said neither yes nor no. Not even a smile. Does he at least understand that we can't pay him?

The following evening, he's walking through our palace.

"A place like this for an election campaign," he grumbles, "it'll cost a lot."

Then he smiles, and, as though he wanted to prolong the fairy tale, he says:

"Okay, I'm willing to rent you what you need to furnish this place at a price that is . . . well, very minimal. But on one condition: you'll pay for any piece of furniture that is damaged."

It may be that my concierge doesn't know anything about me, and it may be that the number of our potential voters can still be counted on the fingers of one hand, but nonetheless, on the eve of the long Christmas vacation, we already have what is surely the most beautiful campaign headquarters in town. A month later, I notice that the headquarters of the Liberal Party, the party of the president of the republic, is not half so elegant as ours.

Then we go off on vacation for two weeks, to recharge our batteries for the months to come. Like Europeans in the month of August, in the month of December Colombians are at the beach. I'm in Cartagena, with Melanie and Lorenzo. Their father has been living in Bogotá for more than three years now; we've moved beyond the consecutive agonies of divorce, and the children have gotten used to living part of the time with Fabrice and part of the time with me. We've been eager for this vacation to begin, as I've barely had time to breathe while I was working at the ministry of foreign trade.

Suddenly, one morning as I'm half-listening to the radio while I prepare to go to the beach, I hear this extraordinary news: "Yesterday was the last day for candidates to register for election to the legislature. The rolls are henceforth closed." Good Lord, what registration? Did we have to register to be candidates? A terrible panic sweeps over me. No, it's impossible, we couldn't have done all that for nothing! We've moved heaven and earth, called upon all these wonderful people, and now we'll have to give up because we didn't fill out a form?

I quickly call Clara. But Clara's on vacation and I don't know where to reach her. Then I'll call the administration in Bogotá. They can tell me. This is incredible. Why didn't anyone tell us, not even Mama? The telephone rings and rings. Why doesn't anyone answer in Bogotá? Why . . .

"Maybe because it's Saturday, Mama."

"Of course, you're right Melanie, you're right. Everything's closed. I'm not thinking, excuse me, but you know . . ."

Melanie doesn't know. She's eight years old and she wants to go to the beach.

I spend one of the worst weekends in my life, imagining all the possible scenarios, imagining our defeat, stalled before we've even begun. It's pathetic. Ridiculous.

Finally Monday comes, and hope is reborn: because of protests here and there, the deadline for registration has been extended for two weeks.

This unforgettable registration proves to be full of lessons. The first thing to do is to get a party to accredit us. In Colombia, you're either a Conservative, like my father, or a Liberal, like my mother. Ideologically, there's not much difference between the two parties. In practice, both have equal numbers of corrupt politicians. So, since Mama is still a leading figure in the Liberal Party and the latter traditionally expresses greater social concern, I choose to seek its endorsement. I imagine that I'll be subjected to rigorous questioning, and I get ready for it. But nothing happens as I expect. The party office is hopelessly confused—people running about in all directions, shouting at each other, laughing, arguing vociferously. Where's the secretary-general's office? The open door over there, on the left. He's there, at the center of a permanent whirlwind. My mother tries to explain, to introduce me to the man, who is crude, odious:

"Okay, okay, here's the accreditation."

He's already talking to someone else. He says neither hello nor goodbye. Then I understand that he gives this accreditation to anyone who asks for it, anyone, without asking any questions, without asking anything about the goals of the candidate who's going to run under the party's banner.

All this seems pretty crazy to me, and yet I'm proud to be a full-fledged member of this group.

The next thing to do is to register as candidates. The following day, Clara and I go together, each flanked by a witness, as is required (Mama is my witness).

Here we see the depth of the abyss that separates us from the other candidates. While we're introducing ourselves, other candidates arrive followed by a regular entourage: supercharged supporters carrying fluttering banners and wearing T-shirts emblazoned with their candidate's picture, fanfares, groups of photographers and camera men. And us? Pitiful! Not even a badge on our lapels! If we needed an electric shock to make us aware of the enormous distance we have to cover, we've gotten it. We've got the message.

Clara and I come out feverish and stunned. It's already mid-January and we don't have a slogan, not a poster. Then I remember a young public relations guy with whom I got along when I was campaigning for Gaviria with Mama. What was his name? Oh, yes, German Medina. I make an appointment with him for the following day. There's no time to lose.

"I'm going to help you, Ingrid, and we'll talk about money later. But where's your program? I can't do anything without your program."

"It can be summed up in three words: Fight against corruption."

"Okay, but you're a candidate in Bogotá, and you need to say what you want for the capital."

"The construction of the subway that we've been awaiting for half a century, the protection of the air we breathe, which is now among the most polluted on the planet, and a policy to help families and children. But none of that will be possible if corruption absorbs half the funds appropriated. Find me something that will symbolize my battle against this disease."

Two days later, German brings us . . . a condom! Clara is scandalized, but I am seduced. At once I see how powerful this symbol is, precisely because it's shocking, because no one can be indifferent to it. Voting for us is like putting on a condom, protecting democracy against the disease that is corruption. It's 1994, in the midst of the spreading AIDS epidemic, and the condom makes the analogy between corruption and AIDS immediately evident. I'm completely won over.

"That's a great idea, German! Terrific! The answer is yes. Go to it!"

He has already sketched out our poster: my photo alongside a picture of a condom, with this slogan: "The best to protect us against corruption."

An idea strikes me.

"You know what? I'm going to hand out condoms . . . in the street. On the model of the campaigns against AIDS. Corruption is the AIDS of Colombia. In addition to the real AIDS, we're suffering from corruption."

I've followed the advice of my mother's friend and called everyone in my address book. Once again, I call these people, most of whom were willing to help me, and say to them: "Do me a favor. Bring me condoms. I need hundreds, thousands of condoms. You'll understand later, trust me."

I stand at traffic lights and knock on drivers' windows.

"My name is Ingrid Betancourt, and I'm a candidate for the legislature. I believe corruption in politics is the equivalent of AIDS. We have to protect our democracy by voting for honest people. Here, I'm giving you this condom so you'll remember me on election day."

People listen to me, some shocked, others troubled, but at the end they do support me. The women say: "What you're saying is true, you're courageous." The men, who are sometimes intimidated and sometimes mocking, say: "Of course, with these things no one is going to forget you."

My father soon finds out about the condoms and things become very tense. He phones me one evening, very upset.

"One of my friends saw you on the street, Ingrid. You don't have the right to do this to me. My own daughter handing out . . . handing out . . . It's disgraceful, degrading . . . How could you? I'm ashamed of you, Ingrid!"

And my mother, from whom I'd hoped to receive some words of comfort:

"Your father is hurt, deeply hurt, and in my heart I understand him. You're campaigning the way your age and your enthusiasm lead you to campaign, but it's disgraceful to distribute those things, what do you expect me to say to you?"

Two days after this little family spat, I'm invited to one of the dinners that are supposed to introduce me to journalists. One of the journalists, a certain Luis Enrique whom I've just hired as my press attaché, has organized this reception. I arrive still shaken by Papa's words. Felipe Lopez, the director of *Semana*, the great Colombian weekly, is there. Does he sense that I'm not in the best of shape? In any case, he holds out his hand to me.

"Well, how's the campaign going?"

"Not too well, to tell the truth. My father's upset that I'm handing out condoms on street corners."

He breaks out laughing. He questions me, comforts me. From his point of view, this is only a generational conflict, since basically he thinks it's an excellent way of attracting Colombians' attention.

The following week, a little note appears under the heading "Confidential," *Semana*'s most widely read column. It says that I'm conducting an offensive, modern campaign that deeply shocks "Ingrid's father, the former minister Gabriel Betancourt."

All at once, the media discover that a young woman of thirty-three, the daughter of a minister, is daring to hand out "rubbers against corruption" in the streets of Bogotá. And everyone rushes to cover the story: television cameramen film me handing out condoms, my picture appears in all the newspapers. I cease to be anonymous, I become a presence. I'm recognized on the street, and I no longer need to knock on the drivers' windows. People open the window and smile at me.

My mother phones me back. "It's incredible, but your father is beginning to find this business with the condoms rather amusing.

Some of our friends are even going so far as to say that it was a smart idea, and that you're making people think."

They thought they'd been dishonored, but now they find themselves admired, and they're secretly proud of my "scandalous" actions.

At this point, something completely unexpected happens: the star of Colombian television, Yamid Amat, invites me to be on the evening news. I'm scared; I know this is my one chance—if I blow it, I'm doomed. I'm expecting the worst, because Yamid Amat is proud of his reputation as an implacable interviewer.

"No one knows you in Colombia," he begins. "Tell us who you are and why you're seeking a seat in the legislature."

"I want to fight corruption."

And at that point I see him obviously stifling a belly laugh.

"Fight against corruption? What are you going to do against corruption?"

"Accuse the corrupt from the rostrum in the legislature."

"I see! Because of course you know people in the legislature who are corrupt?"

"I know a great many of them, yes. And I imagine you do, too."

"True," Yamid Amat replies coolly, "but I won't name names. Are you prepared to name them?"

"Yes."

And then I give five names that occur to me. The names of the five politicians I consider most corrupt.

Yamid Amat is stunned. He pauses for a moment and then, suddenly changing his tone:

"This business with the condoms. Colombians are shocked, you know."

I give my campaign speech and we leave.

Luis Enrique, my press attaché, is enthusiastic about the last part, about the condoms. Since the program is not broadcast live, he's sure

Amat will cut the part where I named names, "which is equivalent to a death sentence for you, Ingrid." We laugh.

I call my parents and all my supporters so they don't miss the evening news that night. Then I sit down in front of the television. And then, the unthinkable: Amat retains only my denunciation of the corrupt politicians. It's a bomb! I myself am astonished, speechless. But not for long: twenty seconds later the telephone starts to ring.

"You were great, Ingrid! Just great! But you're going to get yourself killed. You don't realize that these guys are criminals."

Dozens of laudatory but nervous phone calls, and, amid all this, the deep, solemn voice of a man universally respected in Colombia: Hernán Echavarria, a major industrialist and a former minister of finance:

"Ingrid, I just heard you. I'm with you and want to help you. Do you need money?"

Of course I need money. Right now, we don't have a penny left; everything has been spent on advertising.

"Yes."

"How much do you want?"

"I don't know. Five million."

I think: That's crazy. He's going to hang up on me.

"I'll send you a check tomorrow morning."

At this point in my campaign, the Liberal Party names its candidate for president of the republic: Ernesto Samper. The presidential election is to be held two months after the legislative elections. My mother urges me to resume contact with the Liberal Party headquarters, where I've not set foot since the day I went there to get my accreditation. I've conducted my campaign without any support from the party, without even consulting it regarding my slogans.

"At least let's go say hello to Samper," Mama tells me. "He's a friend, and he'll surely have advice for you."

Samper is going to give me advice? I don't expect anything from him. His cynicism and irresponsibility are fresh in my memory. But I don't refuse. Samper has a good chance of being elected president of the republic, and if I'm also elected, I'll have to work with him.

My mother and I find him in that crazy campaign atmosphere, with overexcited people scurrying about. He's just as he always was: easy, pleasant, carefree, charming. Right from the start, he thinks my bid is doomed:

"You certainly can't think you're going to get elected by handing out condoms! Ingrid, be serious! If that were politics, anybody at all could get into the legislature."

He laughs.

"Ernesto, Ingrid is just starting out. Advise her, please."

"She's learning the ropes, Yolanda, and that's good. Nothing better for learning the trade than falling flat on your face. Help her yourself. Go ahead! I can imagine the two of you . . . Ha, ha, ha! But don't count on me to help you hand out condoms. You know, Yolanda, you're looking younger every day!"

He shows us out, his mind suddenly elsewhere.

Mama is angry, hurt.

"He was odious," she said.

"What did you expect him to say?" I reply. "He obviously thinks I have no chance at all! He can't understand that I'm putting my faith in the Colombian people, in their civic sense, in their desire for democracy. You'll see. Politicians like Garavito, Espinosa Mestre, and others went to jail in trial 8000 because they bought votes for Samper's campaign. This is all they understand. I will show them there is another way."

It starts out badly; it's raining in Bogotá on election day. Rain, as everyone knows, is the enemy of democracy. It discourages independent people from going to the polls, but it has no effect on those who've been paid to vote. They'd go on their knees if they had to. Clara, who was trained as a lawyer, has done a marvelous job of organizing things. We have our people in every polling place, people she has trained to detect fraud and told what legal steps they must take immediately. There are a number of matters to be aware of. You need people at the very beginning, before the polls open, to ensure that the ballot boxes come in empty. Sometimes there are ballots inside already. You have to observe to make sure that people are free to choose their ballot, without interference or pressure. You have to watch that the identity of the voter is real. (We have discovered that dead people have "registered themselves" to vote.) You have to ensure that in the counting, the results are properly registered. Sometimes juries are paid to be blind, while others distort the results. At the end, you can have the correct sum for the ballots but they are assigned to the wrong candidate. Then, you have to be sure that what is being reported to the central national counting office matches with the local information. To guarantee that polls are "pure" and to prevent fraud is a real science.

As for myself, I've chosen to bring sandwiches to all these volunteers, which allows me to buck up their morale and take the pulse of the electorate. In the southern part of the city, it's Mama's children, now grown up, who are watching out for our interests. There, everything is different: they encourage me, hug me, give me an incredibly warm welcome. I'm all the more moved by this because they know they'll get nothing in return for their support, whereas their neighbors who are voting for my opponents have been promised a job, an envelope full of cash, preferential treatment, etc.

Four o'clock; the polls are closing, and I return to our headquarters. Clara is already there. We shut ourselves up in our office, turn on the radio, and collapse, unable to say a word to each other. My parents, who've taken no part in my campaign, have not come. More generally, there's a void around us, as if everyone were waiting at home in order to avoid contaminating others with their fears.

Hundreds of candidates have just fought for months for Bogotá's eighteen seats in the House of Representatives. Clara and I were two minuscule voices in this chorus of smooth talkers prepared to make all kinds of deals, all kinds of arrangements to get into the legislature. Has anyone heard us? That suddenly seems very unlikely. If we weren't so nervous, we'd laugh about it.

At five-thirty, the first projections come in. Names are ticked off, and, to my amazement, my name is there, in fifth place, among the eighteen leading candidates. I think we are screaming—yes, we're standing up and screaming: "It's not possible! It's not possible! They're making fun of us, or they're mistaken. Ingrid Betancourt! Ingrid Betancourt!"

We shout my name over and over, as if it were someone else's. But no, it's mine, it's really mine. The journalist is just as stunned as we are. He says that I'm the surprise of the election, the great surprise, but he's a little short of things to say because he has no information about me. I wasn't among the favorites, and they know next to nothing about me. Who is she? Where did she come from? Then the telephones begin to ring all around us—our supporters, bursting with joy. They laugh, they cry, they come in, bringing everything they have at home to drink. There are other calls, too, dozens of them, and of course the newspapers, which are going to have to describe a woman about whom they know nothing except that she dared to hand out condoms against corruption.

The astonishment turns into delirium when it's announced that I've gotten more votes than any other Liberal Party candidate. It's a miracle! I, who had received nothing from the party, am its most successful candidate in Bogotá. This is the finest victory of my life, for it is the richest in hope. And that's what I tell the journalists that evening: "We've just proved that Colombia is ready to combat corruption. It has chosen the ethical, the democratic, over the venal. It has clearly turned its back on the political class that does not respect it, that has been deceiving and robbing it for decades. The political class that never for an instant believed I would win is now going to have to deal with me." With me alone, unfortunately, for Clara is not elected.

That night, our friends and supporters come in from everywhere. And this extravagant, magnificent headquarters, which we never succeeded in filling during the campaign, is finally overflowing with an exultant crowd. My parents and Astrid arrive, radiant. Finally Fabrice, carrying Lorenzo on one arm so he doesn't get trampled, and holding Melanie with his other hand, comes in.

Four years have gone by since our separation, since my departure from Los Angeles. I took the risk of making my private life miserable in order to regain my Colombian identity. Colombia has just taken me to its bosom, and my friends and relatives are rushing to me, a little dumbfounded by what's happening, but loving, faithful, supportive.

# CHAPTER TEN

I'VE BEEN A REPRESENTATIVE for only a week when the candidate for the presidency of the republic, Ernesto Samper, calls me. He wants to see me right away. His tone is no longer jovial, but it is warm and friendly. On behalf of the Liberal Party, which has not sent me a word of congratulations, he tells me how pleased he is with my victory. I think of Mama; Samper seeks me out even sooner than she could have imagined.

This time, he receives me with full honors. The image of me handing out condoms no longer makes him laugh. Has he sensed a new wind blowing that might be to his disadvantage? Of course. And that's why he needs me, I believe, to provide cover for him on the moral issue. The presidential election is two short months away—that's how much time he has left to prove to Colombians that he is not only a clever politician but also a bearer of moral values.

"Ingrid, the Liberal Party absolutely needs a code of ethics. I'm going to create, within the next twenty-four hours, a commission to draw it up. It goes without saying that I want you to be on it."

I see in this proposal an unexpected opportunity to develop some ideas that I feel are important to the agenda, so I accept. I am well aware that for Samper, my participation is needed only for electoral purposes, but I am hoping to beat him at his own game while the Colombians who elected me watch it happen.

The so-called Commission for Liberal Renewal, a showcase for candidate Samper, consists of ten members selected among the most brilliant, the youngest, or members of the legislature who scored highest in the recent elections. At the first meeting, it's clear to me that these people, delighted and flattered to be there, clearly have no intention of doing any serious work. As is often the case in Colombian politics, the purpose is to announce a program, not to execute it. People believe things are going to change, and then a few years later everybody discovers that the vaunted commission, set up with great fanfare, has not produced a single idea and thus it is necessary to start all over.

At our second meeting, only two of us remain present, a rather reserved young party official and myself. That's not so bad, I figure, we'll lose less time jabbering. As far as I am concerned, I want to begin working.

For a month, I write, article by article, chapter by chapter, a genuine code of ethics. Naturally, I place emphasis on the strict regulation of financing. From my work in the zones where smuggling persists, I've learned that the local elected officials there are financed by the mafia in order to serve its interests. In fact, as I write the code, I am sitting next to representatives who are among the most famous smugglers in Maicao.

From my election campaign, I've learned that some candidates conduct campaigns that cost six hundred million pesos, even if this breaks the electoral law, whereas others, like myself, do play by the rules, spending for the same campaign less than 5 percent of that amount. I ask that all financial records of all candidates accredited by the party be open to public scrutiny. I'm well aware that principles don't count for much in Colombia, so I prescribe draconian sanctions for those who commit fraud, including, in particular, permanent expulsion from the party. In short, according to the terms of my code, only those who are elected using clean, officially declared money will be able to claim membership in the Liberal Party.

Whether this text will be adopted by the commission remains to be seen. As soon as the first draft becomes a matter of interest to the press, the other members of the commission reappear. Not to give a hand, but to block the project. Some are driven by the desire to be in the spotlight, and others by fear, because they are already imagining the sanctions that would be imposed on them if this code becomes the party rule. In any case, a ferocious debate begins.

For two weeks, the commission meets every day, and—*mirabile dictu!*—not a single member is absent. At first, I am the sole defender of a radical ethics agenda, but soon I'm joined by an influential man, Humberto de la Calle, who is supposed to become Ernesto Samper's vice-president. De la Calle is no choirboy when it comes to politics, as I will later discover, but he's capable of great intellectual rigor. In this case, he endorses my concern for strong disclosure measures. Even if he benefits from the system, he thinks it has to be profound-ing reformed, because Colombia is losing its soul, and this code pre-sents an opportunity not to be missed. Does he foresee that this code will rapidly become a weapon, a kind of boomerang, used against the president himself?

Thanks to Humberto de la Calle, my code of ethics is preserved almost intact. Then it is presented to Samper. To my great surprise, although I'm now familiar with his behavior, I see him give his accord without even taking the trouble to skim over the articles.

"Excellent! Perfect!" he cries. "This is exactly what the party needed! The Colombian people . . ."

And slapping his hand down on a text he's not yet glanced at, he launches into a long, grandiloquent speech about confidence, account-ability, and clean politics. There is his new crusade, released just days before the election. Ethics has just become another slogan.

A press conference is scheduled for the following day. The whole administrative staff of the Liberal Party, ministers and former ministers, line up behind the candidate. "Samper launches the Liberal Party's code of ethics," trumpet the headlines. It's clear that the impact on the population is considerable, all the more because the sanctions are specified in scrupulous detail. The journalists report Ernesto Samper's opening speech, but they don't mention that he retreated before the barrage of questions, leaving Humberto de la Calle to answer. Samper himself would have been completely incapable of doing so.

# CHAPTER ELEVEN

**IN THE HOUSE OF REPRESENTATIVES,** my reputation has preceded me. Over there, I am the "bad guy," the traitor, the one who on television named names, pointing out the five most corrupt legislators, the maverick candidate with the condoms campaign, the author of the Liberal Party's ethics code. (The press, showing that it's nobody's fool, will call it "Ingrid's code" just to annoy other members of the House.) And despite taking a hard line on ethics issues—or because of it—I am also the one who has obtained the highest electoral score in the Liberal Party.

One day when I'm sitting alone, as usual, in a corner of the legislative chamber, a rather eccentric-looking guy comes up to me. He's big and friendly, has a mustache, and wears rings on all his fingers.

"Guillermo Martinez Guerra," he says, introducing himself. "I'm a former pilot in the air force, recently elected, like you. People here are strange, aren't they? I've noticed you don't hang out with them much, either. Listen, I'm throwing a little party at my place next weekend. Why don't you come? I'd like to get to know you better. There'll be a couple of independent-minded people like us there."

Yes, I say to myself, I'll go, if only to have two or three people to talk to in the legislative chamber. It's pretty depressing to see how everyone around me, at every session, looks away and discreetly changes seats so as not to be seen near me.

At the party, there are only two other people in addition to Guillermo Martinez Guerra, our host. There's a woman whose strong character—and showy dressing—I've already noticed, Maria Paulina Espinosa; she will become my only friend in the legislature. The other guest is Carlos Alonso Lucio, a man famous for his moral intransigence and for having belonged to M19, the former guerrilla movement that was the most dedicated to establishing a genuine democracy and signed a successful peace agreement with the Colombian government in 1990.

The four of us find ourselves sitting around the lunch table. At one point in the conversation, Guillermo Martinez Guerra mentions a monumental purchase of a special kind of machine gun named "Galil," which was rumored to be almost concluded with Israel. This strange contract to buy obsolete and expensive guns would bring its signatories large payoffs.

"Let's investigate, and if it really involves corruption we'll organize a debate in the House on it," Lucio proposes. "We need to show Colombians as soon as possible that they haven't elected us for nothing, that things are going to change."

We all agree, and we all think we're able to obtain reliable information quickly. Lucio, because of his past membership in the guerrilla movement, is well acquainted with the functioning of the Colombian army. Guillermo Martinez Guerra, a former pilot in the Colombian army, is able to reach the highest levels of the military hierarchy. Maria Paulina, through her husband, who sells helicopters to the air force, and because she is also a reservist, can obtain additional information. Finally, I think I can rely on a friend, Camilo Ángel, who is high up in the arms trade. He and his father are the Colombian representatives for the Colt company. I've known them for a long time. It was at his place that I met Felipe López, the famous journalist who publicized my condom strategy.) I think I can count on him to enlighten me on this deal involving the Galils.

But I'm mistaken. Camilo refuses to give me precise information because Colt was competing against the Galils on the bid and he thinks it wouldn't be correct to disclose information against his competitor, especially after losing the contract against them. He tells me only one thing: "Go to it, dig around—it's a rotten contract, disastrous for Colombia."

Then we demand that the ministry of defense supply us with all the documents relating to this contract. As legislators, we are entitled to obtain this kind of information. These documents show us what's really going on: Colombia is buying an old factory to manufacture Galils, at the full asking price. The seller has obviously found no buyer other than the Colombian army to purchase not the machines, but the obsolete equipment to produce them! Of course, the contract is accompanied by a payment under the table. But that's not all: the Galil is a rifle designed for use in the desert; it's known to jam when it gets wet, and it won't function properly in a tropical climate. To equip our soldiers with this weapon—if we ever manage to produce it—when the guerrillas have modern, well-adapted arms is simply suicidal.

With this information in hand, we hold our first press conference. Journalists rush to attend and the impact is enormous. "Four Representatives Denounce an Extraordinary Case of Corruption" is the headline in the next day's paper. We're nicknamed the "Four Musketeers of anticorruption." Just as we hoped, and as we've promised our electors, we embody a new generation of politicians who are breaking with the venal methods of the old political class.

But our goal is to go further and to strike harder to reveal the truth by organizing a major legislative debate on this issue, to be played out in front of the entire population. Since the Galil contract was negotiated by the preceding government, we're naïve enough to think that Ernesto Samper, who has just been elected head of state, and his appointed ministers will hasten to support us in order to prove to the nation that times have changed. Wasn't it Samper who initiated and then ostentatiously "sold" to the country the Liberal Party's new code of ethics? And to crown our good luck, the new minister of defense, Fernando Botero, is a decent man—at least that's what I believe. He's the son of the painter Botero, my parents' old friend.

Botero receives me warmly.

"Ingrid, I'm scandalized by what you tell me. I knew nothing about it, I'm hearing this for the first time, and it's very serious. I'm going to call for the opening of an investigation immediately. You can count on me. I'm on your side in disclosing this scandal."

The four of us think the great parliamentary debate we're preparing will take place under the best auspices.

But soon, the press, which had been favorable to us, begins to change its tune. It's claimed that Lucio and Guillermo Martinez Guerra are interested in this contract because they're arms traffickers themselves. It's suggested that Maria Paulina's husband is telling her what to do. As for me, my friend, Camilo Ángel, is supposed to have been manipulating me from the outset, in hopes of getting back the contract for Colt. We, who presented ourselves as young crusaders for morality in politics, find ourselves brutally reduced to the level of the politicians we're denouncing, those whose actions are determined solely by their own personal interests.

We're stunned, unable to determine what has led the press to make this turnaround. Later, I learn that the Army Intelligence Service has methodically fed certain journalists these apocryphal stories to discredit our investigation on the orders of the minister, Fernando Botero. At the time, though, we're still innocent enough to imagine that all we have to do is contact these journalists and explain to them their error, and persuade them of our good faith. But none of our explanations seem to convince them. In fact, they produce the opposite effect: the more we try to justify ourselves, to demonstrate and clarify, the guiltier and more pitiful we come across in the articles written about us.

Then, suddenly, the press campaign becomes much more intense. And this time, it's focused exclusively on me, as if the people behind the scenes had understood that by bringing me down as the symbol of anticorruption, they would also bring down our allegations against the "Galil" contract. The press no longer claims I've been manipulated by Camilo Ángel. Instead, they claim that Colt financed my election campaign and that by denouncing the contract to buy the Galils, I'm making a payback to allow Colt to win the bid. Soon, I am accused of having been corrupted myself. The press says that I am the one who is manipulating, using front men, notably my friends, to get payoffs on the deal. And now my friends are being pursued by the press because of me. Thus, in the newspapers, my image as the poor dupe who doesn't understand anything is changed to that of a tricky operator who will do anything to achieve her ends. I am caricatured as Ingrid "Betancolt," and within twenty-four hours I'm attacked by every publication in the country.

To grasp the extent of the media turmoil, one has to imagine the aggression of the press that one day supports the rise of a new moral leader, and the next day furiously attacks the product of their own creation. The journalists—who have been keeping quiet about political corruption for years—felt relieved to think that this person was rotten, just like the rest.

I've had no experience with the press, and I don't know how to manage this kind of defamation campaign. Not only do I agree to talk on the telephone with dozens of journalists who call my home at sunrise, but I also respond at length to all their questions. Since I'm nervous and terribly anxious, I talk too much and too quickly. I am only hoping that my honesty will ultimately shine through.

"Camilo Ángel, he's your friend, right?"

"Of course he's my friend! I'm not going to tell you otherwise."

"His father sells rifles. Did you know that?"

"Absolutely! And Camilo works with him, in case you didn't already know that."

"For Colt! Who lost the contract to the Galils. And after that, you dare to claim that it's a mere accident that, having just been elected, you set out to wage war on the Galils."

"It's a pure coincidence, I swear it."

"There are rumors that Colt financed three-quarters of your election campaign."

"That's false! I haven't received a penny from them. I've published my financial records."

"Listen, you're in a position to know what that kind of publication is worth."

"Precisely, that's why I've campaigned for full disclosure."

"Then why would Camilo Ángel support you?"

"He didn't support me, he didn't finance me. He's a friend, that's all."

"That's not quite enough for explaining yourself, is it?"

# CHAPTER TWELVE

I CAN NO LONGER SLEEP; I'm exhausted. Each new article, by clev-erly exploiting my inexperience and clumsiness, makes me look even more compromised. Yet I have to prepare for the upcoming parlia-mentary debate we have requested. I have to concentrate, go through the arguments, demonstrate that everything we've been saying from the outset is true: that the contract for the Galils is corrupt and that it presents a serious threat to our soldiers' safety.

Then I receive a terrible blow. On its cover, the magazine *Cambio 16* publishes a photo showing Camilo Ángel and me on horseback. And what is Camilo wearing, prominently displayed? A cap with the Colt logo. The article explained that this photo was taken during my election campaign. When I see it, I'm incredulous, petrified. How is this possible? I remember the horseback ride perfectly. One Sunday in the midst of the election campaign, we had called all our friends to come with us on an "ecological" foray in Bogotá. It was perfectly innocent, but the image is disastrous, seeming to confirm all the accu-sations against me. How did this picture get there?

Luis Enrique! My press attaché. He'd offered me his services (which had, in fact, been very useful), explaining that I wouldn't have to pay him until the state reimbursed me for my campaign expenses.

I recall that in the frenzy of my work once the campaign was over, he asked me for the money on two or three occasions. I'd sent him away a bit roughly, perhaps, because the state still hadn't reimbursed me. That's it, he had to be the one. He took hundreds of photos during the campaign, and now he's getting reimbursed by selling them to the media.

I'm distraught, floored. This photo amounts to an assassination. How could he do such a thing? How could someone who'd worked with me, who knew of my integrity, my ideas, my convictions, who knew more than anybody else that Camilo Ángel hadn't participated in the campaign, do that to me? It had to be for money, but he would have gotten the money two months later anyway.

Immediately the phone starts ringing again. The major television evening news program invites me for an interview. For a second I think about refusing, as they've been consistently mean to me, but at the same time it seems like an important opportunity to defend myself. It is impossible to remain silent. I accept, but on one condition: that the interview takes place live. I demand this on a whim, an intuition, without suspecting that I've just taken my first intelligent step in the whole process, during which I've been treated like a hunted animal, making one error after another.

In the hours preceding my appearance, I stay home, deeply depressed, wondering how I will explain to Colombians in a brief, lucid manner that this whole imbroglio is a conspiracy against us. As I think, I leaf through *Semana*,* and I come across a curious article. It reveals that the evening news anchor who has just invited me for the interview is going out with the other female anchor on the show. They've fallen in love; what a charming anecdote. This is exactly what I need, I say to myself. I'm going to use this as an example to explain myself, and Colombians are going to understand. This idea

---

*Semana* is the most important news magazine in Colombia; it is the country's equivalent of *Time* magazine.

gives me an extraordinary energy, for I am preparing a counterattack. I call my parents, who've remained quiet for weeks, overwhelmed by what I'm going through.

"Watch me this evening."

"Oh, Ingrid! You don't realize the harm they're doing you. People no longer believe you, it's a catastrophe! Your image . . ."

"Mama, I'm going to defend myself, watch me, trust me."

The interviewers, Alfredo Vargas and Inés Maria Zabarraín, waste no time in attacking me:

"Ingrid Betancourt, who has been on the front page of every newspaper for days, is with us tonight. In a dramatic development this morning, *Cambio 16* put her on its cover."

The magazine cover appears on the monitor.

"Ms. Betancourt, that's you, isn't it, right in the middle of the election campaign?"

"That's me, yes."

"And this man wearing a cap with the Colt logo on it is Camilo Ángel, isn't it? Not to be malicious, but it seems to me more than just a coincidence. In short, you need to give some credible explanation to the Colombian people, don't you think?"

"I do, but it isn't easy because I have to prove that what seems so obvious is false. I'll give you an example. A short while ago I was reading in *Semana* that you've fallen in love with Inés Maria Zabarraín, who's present here this evening, and that you two are not only working together but also going out. Well, having read this, I might deduce that you're living together. But that would be hasty on my part, because I believe Inés Maria will wait to be married before living with you. And it's the same way with me: You might deduce that because Camilo Ángel is wearing a cap with the Colt logo, I am working with this company, and yet it is not so."

Alfredo Vargas blushes furiously. A commercial comes on before I am even allowed to finish. The journalists are shocked, even outraged. In Colombia, living together outside marriage is very much frowned upon. My attack seems uncalled for—just like the accusations made against me. The newscasters' indignation shows me that Colombians must have finally understood this comparison and what is happening to me.

And in fact, as soon as I get home, supportive telephone calls start coming in.

"We have a clear picture now," says one of my colleagues at the Ministry of Foreign Trade, who calls to support me. "Good for you, you defended yourself like a lion."


It's about ten days before the beginning of the parliamentary debate. Heartened by this initial success, I recover my courage. At this point I receive a decisive helping hand from a man whose aid I'd been hoping for during the preceding weeks. His name is Agustin Arango. He's the Bogotá representative of the French armaments firm Famas, and I've learned that he also played a role in the famous call for offers that resulted in the Galil contract. I've called upon him several times and each time he refused to talk to me. Remaining faithful to a commitment of discretion he made with regard to what could involve the Colombian government, he was unwilling to respond.

"I can't tell you anything . . . Just between the two of us," he told me, "if I talk, I'm a dead man."

The day after my appearance on television, Agustin Arango calls me up.

"Ingrid, what they're doing to you is disgusting. I've thought about it, and I'm going to explain everything to you, show you the documents and give you the names of the guys who are in it up to their necks. But promise you'll never reveal you met me."

I promise. And if I now allow myself to reveal our secret collaboration, it's because some time after the parliamentary debate on the Galils, Agustin Arango died when the private helicopter carrying him crashed. I have good reason to believe that his death was not an accident.

Finally, the debate on the floor of the assembly begins. The defense minister, Fernando Botero, who promised to support me, has in reality orchestrated a grotesque exhibition in support of the Galil. At the appointed hour, we watch a series of uniformed amazons in miniskirts and leather boots almost turning cartwheels before our eyes in order to promote the notorious Israeli rifle. The representatives—delighted by the unexpected show—dissolve into laughter and applause, as if they were cheering girls at a Las Vegas show. What kind of alleged democracy do we live in, I think, that indulges in such disgraceful capers? I'm ashamed for all my fellow citizens, ashamed that we have these representatives, ashamed because twenty high-ranking military officers, their chests covered with medals, are visibly satisfied as they watch this from the best seats in the house. The excitement is at its peak because the media is there, especially the television cameras. What respect can Colombians have for leaders who applaud such a decadent spectacle?

At thirty-three, I'm going to speak for the first time in front of my colleagues. The whole country is waiting to hear this speech. The press has made this debate a duel between Fernando Botero and me, and it has already declared the minister the winner by a wide margin. The pressure has been so intense that I, being a novice, reacted psychosomatically, my anxiety transforming into fever. That day I wake up completely ill, feeling drowsy, my heart beating like hell because of the strong pills I have swallowed. As I watch Botero's army girls perform their grotesque presentation I feel sicker yet.

Sometimes in life there are moments of exceptional intensity, in which one has the clear feeling that one's destiny is at stake. I have this precise feeling then, and a terrible vertigo strikes me when it's my turn to mount the podium.

As a general rule, in the Colombian legislature no one listens; people chat, get up, come and go. It's like a Persian market, so that one has the embarrassing feeling of talking in a void. But this time, as I begin speaking, an impressive silence invades the room.

"It's incredible," I say, "that a minister of the republic, who comes from a prestigious Colombian family, would put the weight of his authority and even compromise his credibility using these pitiful, sequined girls, to back a contract which he knows perfectly well to be crooked. One has to ask oneself why he does it, what personal interest he has in equipping our soldiers with rifles that are not only greatly overpriced but, most importantly, are technically obsolete and will explode in their faces the first time they cross a stream."

For forty-five minutes I offer, with documentary proof of how the government has succeeded in presenting as the most effective weapon available this archaic rifle, for which Colombia has already paid more than it would have cost to buy ultramodern German, French, or American rifles. Brandishing my documents, I accuse the ministry again and again. One could hear a pin drop. I know I'm scoring points. I'm paying back, one by one, the blows that have rained on me for weeks.

That day, without being aware of it, I acquire a reputation as a strong speaker that I've never lost since. From then on, every time I have to intervene before the House, and then later as a senator in front of the Senate, I'm listened to in a silence that is tense, and, often full of hatred.


Have I won? Even though every effort will be made to rob me of this victory, I've succeeded in what was most important to me. I've convinced the journalists, especially one who has slammed me in the past. She comes over to offer her apologies as soon as I go back to my seat. "I hope to have an opportunity someday to redress the wrong that has been done you," she says. "I'm aware that we've let ourselves be completely manipulated."

The newspapers partly right this wrong when they publish portions of the research on which my speech was based. Thus Colombians are finally informed, made to witness a proven case of state corruption. At the same time, they could understand how the state got off incredibly easy, using its influence on the representatives, who have supported Botero almost unanimously. Why didn't they register a vote of no confidence? Why didn't they demand the opening of an investigation? Because they're also corrupt. Have the people understood this? I am hoping they did.

September 1994. Two months later, the inspector of finances confirms a posteriori my accusations.* He opens an investigation of the parties who signed the contract for the Galils. The scandal can no longer be covered up. The factory, which was paid for and delivered, has not produced a single gun, and never will. It's a monumental mess.

"We've been able to uncover the existence of fraudulent conduct on the part of three high officials," the inspector asserts. He gives the names of these three men, who are to be the scapegoats. As is often the case in Colombia, this maneuver will do away with a scandal without having to prosecute the highest officials responsible for the bribe. The real culprits are too important, and hence, untouchable.

Once again the press evades its responsibilities. There isn't a single acknowledgment that we, "the Four Musketeers," the ones who fought alone to dismantle this rotten net of corrupted officials, the ones who were attacked so as to be silenced—not a single recognition

---

*The inspector of finances is the highest authority in Colombia responsible for watching over the national budget and making sure that the spending is correct and legal in order to prevent the National Treasury from being robbed.

that we were right. Not one single article expresses astonishment at the legislators' cowardly blindness. Not a single word stigmatizes this government, these military men, who are so deeply implicated. When the journalists rush to his office, Fernando Botero acts surprised, haughtily draping himself in the wounded dignity of the state. "It is essential that the judicial system be allowed to do its work," he has the gall to say. "Let it identify the guilty parties and punish them in proportion to their responsibilities."

But the contract is honored and the bribes handed out. Worse yet, the same judicial system that is so quick to cover up a state scandal opens an investigation into my activities on the basis of anonymous letters addressed to the prosecutor. Journalists give me this news with a kind of delighted curiosity, waiting for my reaction. I move again to the frontline—by a strange coincidence—on precisely the day that the Galil affair is buried.

These big front-page newspaper headlines announcing that I'm under investigation are a new, crushing blow for me. For the second time, I feel the extent of the power at the disposal of a corrupt state— the power to annihilate anyone who gets in its way. I've seen what this kind of state is capable of, and I'm truly afraid. It is the first time in my life that I have to deal with the judiciary. My friend, Maria Paulina Espinosa, is also very worried about me. She sends me a man who will henceforth be at my side every time the judicial system tries to take me down: Hugo Escobar Sierra, an elderly attorney and a former minister of justice who knows the ins and outs of a system of which he is a part.

When I meet with him, I'm beaten down, so depressed that I am beginning to feel guilty for being friends with an arms merchant. Is that kind of social connection forbidden once you're elected? Can I be prosecuted and found guilty? Maybe. I don't know. I don't know the law.

"My child," he says to me, "you don't realize what a monster you've challenged. They know they don't have anything on you, but they will stop at nothing to discredit you. The problem with honest people like yourself is that you're always inclined to feel responsible. You've done nothing wrong. You've got to have confidence in me, but especially you've got to have confidence in yourself."

He accompanies me to each of my depositions, and I notice that his presence intimidates the officials. I begin to regain my courage.

The investigation of the Galil affair lasted two months, but the investigation against me will last more than a year. Hugo Escobar Sierra constantly holds my hand, attentive, alert, ready to ward off all the low blows. He refuses even minimal payment for his services. "I'm defending you because I want to defend you," he repeats. "That's all."

Finally, I hear the official in charge of investigating my case utter these incredible words: "We are going to close the investigation of you because we haven't found anything. I'm going to give you a certificate stating that the case is closed. But on one condition: you say nothing about it to the press. Remember that we can always reopen an investigation whenever we like."*

This is barely veiled blackmail. That very evening, I send a copy of the certificate to all the national newspapers expecting this would be a scoop for them. Not a word of it was to be printed by a press whose independence is praised throughout the world.

---

*My "accusor" is an official working under the command of Orlando Vasquez Velasquez, the "Contralor," or national inspector of finance, who himself will end up in prison after trial 8000 for taking money from the Cali cartel.

# CHAPTER THIRTEEN

**ERNESTO SAMPER** is elected president of the republic on June 19, 1994, two months after my election to the House of Representatives. I've supported Samper—notably by agreeing to write his Liberal Party's code of ethics—despite my distrust of him. I want to believe that his talk about ethics and morality is sincere. I endorse the program of the Liberal Party, which is more socially oriented and expresses more concern for social issues than the platform of the Conservative Party defended by Samper's unfortunate rival, Andrés Pastrana.

A scandal involving Samper explodes like a bomb on June 21, two days after his election to the presidency. It is Andrés Pastrana, the defeated candidate, who lights the fuse. Acknowledging his defeat, he cries out: "I have only one question for Samper: Can he swear on his honor, before the Colombian people, that he hasn't received money from the drug traffickers to finance his campaign?"

Samper does not respond to the challenge, but at the same time the press publishes an astonishing document: the official transcript of a tape on which the Rodriguez brothers, the infamous heads of the Cali drug cartel, talk about Samper and reveal that they are investing millions of dollars in his campaign.

Later, it will be revealed that this "narcotape," as Columbians will call it, was recorded by American agents from the Drug Enforcement Administration (DEA) during the election campaign and was given to Pastrana a week before the election.* Evidently, the purpose was to have him use it against Samper. But Pastrana did not divulge it until after he'd been beaten.

The information does not, however, produce its intended effect. Because it is associated with the spiteful act by a defeated candidate? Not only that. In this spring of 1994, Colombians are just emerging from the bombing campaign conducted by the most notorious mafioso, Pablo Escobar, the dreaded boss of the Medellín cartel. Escobar, responsible for hundreds of fatal attacks and for the death of Luis Carlos Galán, has just been gunned down himself, after being on the run for months. The country, wanting to believe a bloody page in its history has been turned, that the mafiosi will never again impose their will on our elected officials, breathes freely again.

Now the country is being told that its new president, even before he's moved into the presidential residence at Nariño palace, may have been financed by the new kings of cocaine, the Rodriguez brothers. Does this revelation mean everything is going to start again? It's too much for people to bear. They simply demand the right to hope, the right to comfort themselves with illusions. And they do so by turning their eyes chastely away from the prophecies of misfortune that are all over the newspapers.

For better or for worse, instead of demanding explanations from the presumed offender, Ernesto Samper, the newspapers attack the person who launched the scandal, Andrés Pastrana. Accused of tarnishing the allegedly restored image of Colombia, Pastrana is soon nicknamed *el Sapo*, the toad, and caricatured in the newspapers and

---

*The press made this revelation by presenting it as "rumors." It was said that the tape was given to Pastrana by the DEA. The U.S. embassy never denied it.

on the city's walls. Vilified and accused, Pastrana—who will be elected president of the republic four years later, in 1998—chooses to leave Colombia in order to escape the resentment of his own people.

Ernesto Samper remains silent. Why would he talk about this shady affair with the public that only wants to believe in him? To a public that's already on his side? On July 15, however, sensing that a storm may be brewing in the United States, Samper takes a bold step: he sends a letter to the American senators—not to the Colombians—to explain that this narcotape is the result of a plot hatched by the Cali cartel to destabilize his government. He asks the senators to help him fight the drug traffickers. This step is cynical but extremely intelligent. It has two effects: on one hand, Samper, by allying himself with the U.S. senators, embarrasses the U.S. government, and on the other hand, he is proving to the Colombians his good faith. Furthermore, it conveys to us, Colombians, this implicit message: the Cali cartel is after my hide, just as the Medellín cartel was after Galán's.

I am taken in, just like we all are. I can't believe Samper is capable of making up such a story. It is because I believe his government wants to be perceived as battling corruption that I'm completely disconcerted by Botero's position on the Galil affair. I remember a dinner conversation (a conversation that probably took place all over the country at the time) with my parents about this affair of the narcotape. I recall Papa saying, "It's made up of whole cloth, this affair. Who would accept funds from the mafia these days? Samper's much too clever to have done something like that. Obviously, it's a trick on the part of the Rodriguez brothers. I don't like Samper, I would have preferred Pastrana, but on this issue he's handling things smartly. Since the American government suspects the Colombians of every vice, it's prudent to appeal to the senators so that the Americans will know what is happening."

At that point, we all become the victims of a little flush of nationalism, hostile to the suspicions of the Clinton administration.

To us, they seem to indicate a scorn for our new president and an imperialistic attitude toward our country.

On August 16 the nation's public prosecutor, Gustavo de Greiff, declares, after going over the remarks attributed to the Rodriguez brothers, that there is no reason to open an investigation of Samper. I will figure out, later on, why de Greiff, who knows he's on the way out, has made this decision so quickly, exactly forty-eight hours before the arrival of his successor, Alfonso Valdivieso. For the time being, I suspect nothing, and when later, in August, I receive an invitation from Ernesto Samper to meet with him in the Nariño palace, I am rather impatient and eager to see him again. It is said that the office shapes the officeholder. Has the presidency transformed him as well?

Nothing's changed. He is charming, relaxed, and always ready to laugh. I immediately have the feeling he's interested in everything except the country's problems. We don't mention the very hot Galil affair, on which I'm working, or the notorious tape scandal, except perhaps in the form of a caustic joke characteristic of Samper: "Don't talk so loud, Ingrid, the gringos have bugged my office."

After a few trivial remarks, he asks me for news about my parents. It all sounds innocent; he seems genuinely concerned for their welfare. But this seemingly innocuous conversation will later take on an unexpected importance, bringing me very close to an indictment that threatens to end my political life.

"Tell me how Yolanda is doing."

"Mama's fine. I think she's not unhappy to have finally turned her back on politics. She's devoting her time to the *Albergue*, to her children."*

"Fine, fine. And your father?"

---

*Albergue*—which means "shelter"—is what my mother named her foundation, which rescued abandoned, homeless children. My mother was eighteen years old when she founded the *Albergue,* and since then, over forty years, she has given a roof and provided a family to more than 10,000 kids (see photo insert).

"He's all right. Papa has money problems, but nothing serious. We were talking about it the other day. It's hard to believe, but his pension hasn't been raised in more than twenty years. He said to me, 'I've worked all my life, and look what I've got to live on, three pesos.' When you think that he was an ambassador and a minister, it's absolutely grotesque! Just imagine how it must be for other people."

"Yes, we'll have to look into all that. We'll see, but if I can do something for your father, I'd be delighted to do it."

The conversation ends there. I don't ask anything for Papa, obviously, and don't see anything other than polite attention in what Samper says. As I leave, I learn he has reserved all his afternoons for meetings with members of the legislature. I tell myself he's right to buck up his majority's morale at the beginning of his term. I go back to work on the Galil affair. The parliamentary debate is approaching, and is to be the heart of the political storm.


At the end of November 1994, we've finished with the Israeli rifles, and my three colleagues and I have finally gained credibility in the eyes of the country. The public now recognizes that we're on an important track. The first step in our campaign against corruption is to change mind-sets at the highest levels of government.

Why do we decide to focus next on the concealed and sinister bonds linking drug traffickers with politicians? Because these bonds are eating away at Colombian society. Nothing new can be undertaken until this disease is purged. The Colombian people's reaction to the cassette affair shows how painful this wound is. We would rather point the finger at Pastrana for exposing it than condemn Samper for possibly feeding it. What about the rumors circulating everywhere about the murky dealings between Pablo Escobar and former President Gaviria, before he decided to chase and finally kill Escobar? What about all that is publicly known but not accepted by our people because we choose to harbor hope that these rumors are false, while knowing that if we find out they are true, we would have to face the fact that Colombia is losing her soul. In short, to us, seeking out the truth becomes an indispensable preliminary to the political renewal we are seeking.

As a first step, we decide to organize a legislative debate on national security. This is one way to start a public discussion of the terrorism generated by Pablo Escobar, to ask formally how all that could have happened, and to figure out how the government managed the situation. The Colombian people need to know the truth.

What the truth is we first discover in preparing for the debate. We learn that Escobar had Galán assassinated in 1989 because Galán supported the extradition of drug traffickers to the United States. If Galán were elected president of the republic, the extradition agreement would be adopted and exercised. Gaviria, who presented himself as Galán's heir and got himself elected to replace him, immediately abandoned the extradition agreement. Doing so allowed Gaviria to arrange for Escobar's surrender and put an end to his reign of terror that involved bombings, kidnappings, and assassinations.* Gaviria was credited with a great victory, whereas in fact the state had just capitulated to Escobar.

Why "capitulated"? Because Escobar was not at all the prisoner people imagined. He lived in a luxurious prison known as "the cathedral," surrounded by his staff of about twenty people. More than ever, he was involved in the international cocaine trade. In fact, in "prison,"

---

*Escobar placed bombs inside shopping malls, brutally killing women and children. He killed hundreds of policemen and promised rewards to anyone who would bring him the head of any policeman. He killed the minister of justice, Rodrigo Lara Bonilla. He killed judges, and he kidnapped Andrés Pastrana, and Diana Turbay, the daughter of former President Julio Cesar Turbay. He tortured to obtain information. He would force women to have sex with him and his lieutenants and would kill them afterwards.

his situation improved. Having made bitter enemies of his former allies, the Rodriguez brothers of the Cali cartel, Escobar found safety within the walls of the "cathedral," under the "protection" of the government police. I use the word "protection" because we discovered that Escobar had the keys to his so-called prison and, even worse, that he actually owned it. It had been built for him with his authorization and his money, on land that belonged to him.

We also found out why Escobar escaped from his gilded prison. Not only was he carrying on his cocaine export business from inside, he was also enforcing his criminal rules, handing out punishments to mafiosi, and even pronouncing death sentences. It came to the attention of the prosecutor, Gustavo de Greiff, that a massacre had taken place inside the "cathedral" compound. An investigation confirmed that two of Escobar's men, the Galiano brothers, who were accused of stealing millions of dollars from him, were killed in a particularly vicious way: they were cut up alive with a power saw, barbecued, and fed to the dogs so that no trace of their bodies would be found. Horrified, and certainly concerned about the reputation of the judicial system, the prosecutor asked President Gaviria to have Pablo Escobar transferred to a real prison. But Escobar had made a deal with Gaviria: he had surrendered on condition that he would be detained in the "cathedral." Gaviria knew that if he betrayed their secret pact and sent Escobar to a real prison, Escobar would take his revenge. So, Gaviria managed to let Escobar know that he was about to be transferred, implicitly encouraging him to escape—which Escobar did without the slightest difficulty.

What happened afterwards that allowed Escobar to be killed a few months later? After the debate, we found the answer, thanks to an unexpected encounter with the Rodriguez brothers.

On the evening of the debate, the representatives decide to set up a "National Security Commission" whose first task is to pursue the investigation into the Escobar affair. Since most of the legislators are

Samper supporters, they see this as an opportunity to do in Gaviria and at the same time divert the growing suspicions about Samper. We "musketeers" are appointed part of this ten-member commission.

In February 1995, while attending a working meeting at the prefecture in Cali, we are asked to come outside for a minute. There, a man introduces himself as someone who has been sent by the Rodriguez brothers. They've been following the debate in the legislature, he says. They've heard what we said about the conditions under which Escobar was detained, and they have other information to give us. Would we agree to meet with them? We have to decide right away. There is only one condition: we have to go with this man immediately, without telling anyone. There we are, Lucio, myself, and the third member of the commission, looking at each other as if trying to decide what to do.

"Okay, let's go."

They put us in a car with covered windows. We can't see anything outside. For an hour, the driver seems to be driving as if to confuse us. When we're finally asked to get out, we find ourselves in the underground parking lot of an apartment building. We take the elevator upstairs, then we are led into an apartment. Although it is carefully furnished, it's clear no one lives there. A black servant, her collar and apron neatly starched and ironed as if she worked in a mansion, asks us what we'd like to drink.

Hours pass, the day fades away, and still nothing happens. The only other human presence with us in the apartment is this smiling, attentive woman who regularly brings us things to eat and drink. Are we scared? No. Yes, perhaps . . . But, having been so battered and abused in the Galil affair, aware of the way those in power trapped me by fabricating the "Colt affair"—I'm afraid of falling into another trap. Who is to say we're not going to be taped or filmed talking with the Rodriguez brothers, and that these videos or tapes are not going to be shown on the evening news? Wouldn't that be the best way to silence the people exposing the connections between politicians and the drug traffickers? I mention my fears to the others: "As soon as we get out of here, we're telling everything to the press. Only complete openness can protect us against future traps."

Finally, late in the evening, something begins to happen. We hear the sounds of voices and of doors opening and closing. Our hostess asks us to follow her, and she shows us into a small room furnished with a table and six chairs, three on each side. She asks to sit down on one side of the table, facing the three empty chairs on the other side.

Another quarter hour goes by. Then they come in, and my heart starts beating faster. I recognize these faces, which have so often appeared in the newspapers as the symbols of horror itself.

First, Gilberto, the elder brother, short, with a mane of white hair, comes smiling straight toward me. He offers me his hand and greets me as "*Doctora*," a sign of respect, of deference.

"Good evening, *Doctora*!"

Then comes Miguel, the younger and more slender brother.

Finally, José Santacruz, an icebox of a man, the third associate at the head of the Cali cartel appears.

Gilberto Rodriguez notices our surprise and perhaps also our discomfort. His first words are intended to calm us.

"You're surprised to see us in the flesh, but what really surprises you is seeing that we're normal people. Look at us, we're not wearing jewels or gold chains."

I think to myself: he's right, you'd think they were ordinary businessmen. Open shirts, navy blue pants, moccasins—the way people dress in Cali, the city of eternal spring. Nothing that strikes the eye.

"Why did you want to meet with us?" Lucio interrupts.

Gilberto launches into an amazing speech in response. He goes through all the good deeds they are doing for Colombia, the dozens of legal businesses they've started, giving work to half the city. He goes on to say how they're being unfairly persecuted by the judges considering their only goal is to contribute to the happiness and prosperity of the Colombian people. This is too much. I am beginning to feel out-raged. How dare these gangsters, these criminals, present themselves as Robin Hoods?

"Do you realize that because of you we can no longer travel abroad without being immediately suspected of being drug traffick-ers?" I ask nervously. "You've ruined Colombia's international image, and as far as your good work goes, you've thrown the people into ter-ror and instability. Because of you, Colombians think they no longer have a future."

At these words, Miguel's face goes scarlet. He violently pushes back his chair, grumbles a few unintelligible words, as if he could no longer control himself, and leaves the room. My two colleagues sit there in dismayed silence, as if to say: "Great, Ingrid, we're off to a good start. Now what do we do?"

But Gilberto begins to speak again, calmly, as if his brother's exit wasn't something to worry about.

"Now, we also want you to know what our agreement was with former President Gaviria regarding Pablo Escobar's death."

For the Rodriguez brothers, Escobar is the devil incarnate. When Gilberto talks about him, there is fear in his voice, as if Escobar was still alive. Gilbert is hardly a choirboy himself. Allies for a long time, they started fighting when Escobar asked the Rodriguez brothers to hand over to him one of their associates. Apparently, this Cali cartel chief had not given Escobar his "contribution" and was therefore condemned to death. Outraged, the Rodriguez brothers refused.

Then it was all-out war.

When Escobar (while living in the "cathedral") killed the Galiano brothers with a power saw, the terrified Galiano family went to the Rodriguez brothers for protection. It was Gilberto who convinced them to tell de Greiff, the public prosecutor, about the killing. After

Pablo Escobar's incredible escape, de Greiff assembled around a table—the *mesa del diablo*, the devil's table—everyone who had an interest in having Escobar quickly eliminated. These included a representative of President Gaviria, who lived in fear that Escobar would reveal the agreement regarding his surrender and "imprisonment" in the "cathedral," the Rodriguez brothers, who were afraid that Escobar would have them assassinated, and, finally, members of the Colombian police, who nurtured a deep hatred for Escobar and his *sicarios*, responsible for the killings of many policemen. Thus some representatives of the government sat down with ringleaders of organized crime. Of all the shady deals I believe have taken place, this meeting best symbolizes how sick Columbia is, how profoundly infected by the mafia.

At the conclusion of this meeting, the Rodriguez brothers agreed to find Escobar and point him out to elite police snipers. A dozen men were selected for this killing, and the Rodriguez brothers promised each of them a million dollars in reward. They kept their promises. Gilberto tells us they spent a fortune tracking down Escobar, and, in particular, they financed the development of a technique of electronic surveillance using a phone-tapping system. The Rodriguez brothers knew that Escobar, a monster who had showed his fourteen-year-old son how to dig out a victim's eyeball with a red-hot spoon, passionately loved his youngest daughter, Manuella. Convinced that he would try to telephone her, they set everything up to intercept this call. Escobar fell into the trap. Located in December 1993, he was shot down while trying to escape over the roof of the house where he'd been hiding.

Escobar's fall, celebrated in the whole country and on the front pages of the newspapers, was credited to President Gaviria and his police. My colleagues and I discover to what extent, once again, we Colombians have been deceived and manipulated. We owe this victory over the most fearsome of the mafiosi not to our institutions, but to other mafiosi. And this disease continues its long-term goal of annihilation, for obviously the Rodriguez brothers have exacted a high price for the help they provided. That's what we understand in the course of our conversation with Gilberto, his high-strung brother Miguel (who has finally come back to the table), and José Santacruz.

Multibillionaires, the Rodriguez brothers needed to craft for themselves another virginity and launder their dirty money in order to hand it down cleanly to their children. This involved a surrender, followed by sentencing, and it was the conditions of this so-called subjection to the law that had to be negotiated. Like Escobar, who surrendered only after the principle of extradition had been removed from our constitution, and only in exchange for the promise of incarceration in his own three-star hotel, the Rodriguez brothers have also negotiated with Gaviria, through the public prosecutor, Gustavo de Greiff. Gilberto, the elder Rodriguez brother, was ready to give himself up to allow all his relatives to live a life of wealth, holding their heads up high, over the coming decades. So this is the deal with Gaviria. But, wait a second, Gaviria is approaching the end of his term, and he cannot be reelected. What if . . .

Suddenly, at this point of our conversation, the narcotape affair pops up in my mind. In the recordings, we hear the Rodriguez brothers speaking in very warm terms about Ernesto Samper, Gaviria's successor. At this precise moment I have the intuition that the Rodriguez brothers could have contacted Samper in the event that their surrender could not be arranged for the final months of Gaviria's term. And as if by mere coincidence, it is de Greiff, the same prosecutor seated at the devil's table, who declares there are no grounds for opening an investigation into the Cali cartel funds donated to Samper's campaign. Suddenly everything seems terribly clear to me, and, taking the risk of upsetting Miguel again, I burst out:

"Fine, and how much did you give Samper for his campaign?"

"Humph! Twelve billion pesos," Miguel snaps back, with an arrogant look on his face.

Then Gilberto, embarrassed and gauging the importance of Miguel's blunder, adds:

"That's correct, but Samper didn't know about it, it didn't go through him, he never knew anything."

I smile incredulously.

"Excuse me, but that's hard to believe. When you give money to a candidate the point is that he pays you back when he's elected, isn't it?"

Gilberto pretends that his honor has been wounded. "*Doctora*," he stiffly replies, "we also have the right, don't we, to have political convictions? Lots of people give money anonymously to this or that candidate; why shouldn't we do the same?"

At this point, someone knocks on the door and we're asked to go back into the living room where we waited before; apparently, while the three bosses of the Cali cartel have visitors. We hear the sound of footsteps and muffled conversations, and then, suddenly, while all these men are standing behind the double frosted-glass doors of the living room, we're stupefied to see that the Rodriguez brothers' visitors are policemen in uniform. Are these men really being tracked down by all of the country's police, or is that only a sham?

When we resume our conversation, I say in astonishment:

"You were just saying that you were hounded by the police, but the police seem to get on rather well with you."

Gilberto hints that he, in fact, controls a considerable portion of the police force. "I have good connections," he says. As we seem stunned to hear this, he continues with a certain tone of self-importance in his voice:

"It's exactly the same with the parliament! Most of your fellow representatives are in our pay."

"What do you mean, most?" I say, thunderstruck.

"About a hundred representatives and more than half of the senators, *Doctora*. Would you like their names?"

Though I don't say anything in reply, he names a dozen legislators. And there I am thinking: "If more than half the representatives are on their side (there are 186 representatives and 100 senators in all), they're governing the country more than the president."

On that note, the meeting comes to an end. Over the following hours it occurs to me that the Rodriguez brothers have made an overture to us: Why not join forces with them, since they control the whole apparatus of government? Why should we go on fighting them, fighting corruption?

Two days later, as agreed, we tell the press what the Rodriguez brothers have revealed to us concerning their involvement in Escobar's liquidation. Then we proposed to strongly back up the idea of creating an International Court of Justice to deal with the impunity of Colombian drug traffickers. But we keep to ourselves the main lesson we've drawn from this encounter: the mafia's control over all the nation's institutions, from the Congress—the source of the country's laws—the judicial system, and the police, which are charged with enforcing these laws. This will have an effect on my future thought and action. More than ever, it seems to me that only extradition can halt this fatal vortex into which Colombia is being sucked. Luis Carlos Galán knew that and he was right; he fought for extradition, and paid for it with his life.

As for revealing the Rodriguez brothers' admission that they have financed Samper's campaign, it would be a real bombshell in Colombia—it would be irresponsible for us to mention it. And we have to avoid being manipulated. On one hand, the admission may be, as Samper claims, a strategy to weaken his government (though Gilberto seemed more in the mood to protect him). On the other hand, we have to investigate. If proof exists, we will have to find it first. Only then we will be able to engage in the battle.

# CHAPTER FOURTEEN

MARCH 1, 1995—that is, a few days after our meeting with the three bosses of the Cali cartel—the United States grants President Samper's Colombia only a "conditional certification,"* pending more convincing action against the drug traffickers. This certification determines whether Colombia will receive aid from the United States, and how much.

This is clearly the second warning given to President Samper, eight months after the content of the narcotape was unveiled. But Colombians, who want to believe in their president's integrity, interpret the message as a new affront on the part of the gringos: What right do they have to judge us when they're the main consumers of drugs, whereas we're fighting the drug traffickers and paying for it

---

*This is a measure taken by the U.S. government to ensure that countries that receive financial support from the United States take an active part in the war on drugs. A country perceived as cooperative is "certified" by the U.S. government. A country with a "conditional certification" status is one that is perceived as not being sufficiently productive, meaning that the U.S. government is not satisfied with the results of its war on drugs. A country is decertified when it is perceived to be doing nothing to fight drugs. Much like a "good behavior" certificate, this is used as a tool to pressure foreign governments, and it is considered to be "imperialistic" by Latin American countries.

with our own blood? A flood of anti-Yankee nationalism washes over the country, aided and amplified by the screaming headlines in the newspapers.

By an amazing coincidence, however, on the next day, March 2, Jorge Eliécer Rodriguez, the younger brother of Gilberto and Miguel, is captured by the police. No doubt Samper wanted to offer a hostage to the United States and show the public that it couldn't have hoped for more. The newspapers love it: the Americans accuse us of not doing enough to fight the narcotics dealers, and you see what happens? That'll show 'em! It's very strange, I tell myself, thinking about the good relations with the police acknowledged by the elder Rodriguez. This turn is a little too convenient to be a mere accident, but, it could be, why not . . .

Beneath all this agitation, a judicial land mine directed against President Samper is about to explode and neither the Colombian people nor our anticorruption group have yet noticed it. Remember I said that in mid-August 1994 the public prosecutor had to yield his office to a certain Alfonso Valdivieso? Well, Valdivieso is Luis Carlos Galán's cousin, and he's what we call a "hawk" on moral issues, committed to fighting corruption. While searching one of the properties of the Cali cartel chiefs, a group of officials from the Army intelligence service working in collaboration with his staff come across a list of people who have benefited from the largesse of the Rodriguez clan. At the end of 1994, Alfonso Valdivieso discreetly opens an investigation.

On January 30, 1995, the magazine *Cambio 16*, taking advantage of a leak in the investigation, publishes a list of the political leaders who've received T-shirts and cash donations for their campaigns that were paid for by the Cali cartel, that is, by the Rodriguez brothers. Well, well, people say to themselves, this Alfonso Valdivieso is actually doing something.

He was working, yes, and this T-shirt incident revealed in the press is almost a detail in the broader investigation Valdivieso's men were conducting in silence. We will later learn that it has been discreetly monitored by the United States. The bank accounts of the fortunate recipients of T-shirts are examined by policemen. And, as if by accident, these accounts contain considerable sums, quite out of proportion to the income their owners have declared. Everything is there, in hard numbers. These elected officials don't bother to conceal anything because they're not used to the judicial system actually doing its work. They have an enormous sense of their impunity. Even when the investigators ask them where the hundred fifty million pesos in their accounts came from, they think they can get out of it by simply claiming to have sold a work of art. Our leaders never sold so many works of art as they did at the beginning of 1995.

On April 21, already in possession of documents that could destroy the Colombian political class, Alfonso Valdivieso officially opens what will be known as "trial 8000." (In Colombia, trials are numbered, and it is likely that Valdivieso, sensing that this trial would go down in history, arranged things so it would have a round number.) The prosecutor has already imprisoned one of the prominent leaders of the Liberal Party, Eduardo Mestre, and the list of legislators who are likely to join him in jail contains at least a dozen names.

The announcement of this trial arouses a great deal of excitement in the country. The judicial system has never before bothered with the country's political class. People wonder what to make of this sudden change, particularly since Valdivieso seems not to support Samper, at least based on the fact that most of the elected officials accused by Valdivieso are close to the president. What could Valdivieso's secret goal be? Is he really acting only on the basis of moral principles? People get lost in hypotheses. They have a hard time understanding that a man who's been shaped by the system could suddenly go to war, all by himself, against the system's corruption. I later find the solution to this puzzle: moved by his own presidential ambitions, Alfonso Valdivieso is seeking, by his integrity and his incontestable courage, the support of the United States. To put it more crudely, he's going beyond their expectations.

On June 9, 1995, just after Valdivieso has imprisoned another group of politicians, a thunderbolt strikes: Gilberto Rodriguez has been tracked down and arrested. For me, this is the beginning of an awareness of what is happening behind the scene, which will gradually lead me to get deeply involved in the coming tragedy. The government has actually "arrested" Gilberto Rodriguez, the man whom was invited to the *mesa del diablo*, the man who has the Cali police under his heel, the man who finances half the legislators? That makes no sense at all! This time, I know too much to be taken in, as most Colombians are. Gilberto Rodriguez was not hiding, he was not even tracked by the police. In fact, he was negotiating with the president's office his surrender in terms favorable to him and his family. No, I'm convinced they haven't arrested him, I am sure he gave himself up. Why? To give Samper a lift, of course! At the same time that Valdivieso is decimating the president's political support, that is, arresting his own friends, what does the president do in the face of approaching peril? He provides himself with a spectacular victory: his police finally neutralize the most fearsome mafioso next to Escobar. Who would now dare to claim that the president has received money from Gilberto Rodriguez? Legislators may have gotten money from Rodriguez, perhaps, but not the president. Just look, here's the proof, he's throwing old Gilberto in prison!

That's when I decide to note down every new event with care, so that I can later reconstitute the development of a scandal whose proportions I already sense will be monumental. I begin with these two "coincidences" that are only too easy to see through: the day after the United States issues its nonsatisfactory conditional certification, Jorge Eliécer Rodriguez is "arrested"; the day after the opening of trial 8000, which linked high-ranking politicians with drug dealers, Gilberto Rodriguez is "arrested." What will be the next move in this strange game of Ping-Pong?

The pressure of the United States in the trial of prominent legislators is evident in President Samper's unease. In the spring of 1995, the mood in Colombia grows significantly darker. I'm not the only one to perceive a terrible storm brewing. Everywhere, people seem to be holding their breath and watching for the first signs of the breaking storm.

In this context of disillusionment and anxiety the magazine *Semana* reveals, on June 22, that a military coup d'état is in the offing. People were expecting the worst; would this be it? In any case, the announcement reveals how fragile the Samper administration is after only a year in office.

And, as I've expected, the Ping-Pong game starts up again: in response to this rumor about a coup d'état, José Santacruz, whom I'd met along with the Rodriguez brothers, is also "arrested" on July 4, while having a leisurely lunch in a well-known restaurant. Again, he just wanted to show the public, the military, and the Americans that he was at the helm, that Samper would not have acted otherwise. But what is becoming absolutely clear to me is that the leaders of the Cali cartel are helping Ernesto Samper by allowing themselves to be arrested by the police.

On July 26, however, the scandal takes a threatening turn for the president: Alfonso Valdivieso indicts and imprisons the treasurer of Samper's campaign, Santiago Medina. The news sets the country on fire. Medina has acknowledged that he received considerable sums from the Cali cartel, with the consent of Ernesto Samper and his right-hand man, the current minister of defense, Fernando Botero.

His testimony has not yet been entirely revealed to the public, but the gravity of the situation is implied by the announcement that the head of state is making to the country that same evening.

Samper appears on television in a completely different guise. He no longer wears the charming smile so familiar to me. He's puffy and tense, and at times he has a haggard look in his eyes. In essence, he says that if mafia money infiltrated his campaign, it happened behind his back, through people who betrayed his confidence. A year earlier, he was claiming that this accusation was completely false, that it was a plot on the part of the Rodriguez brothers to destabilize his government. Now he acknowledges that it may be true, but asserts he knew nothing about it.

On July 28, two days after this first temblor at the highest level of government, the House of Representatives, where Samper's supporters are in the majority, takes what seems to be a paradoxical step: it sets up a commission to investigate the president of the republic. In legal terms, this commission is not intended to judge the chief of state but, rather, to say whether or not there are grounds, on the basis of the results of their investigation, to begin judicial proceedings against him. The president of the republic enjoys an immunity that prevents someone like Alfonso Valdivieso from doing anything against him. But if the legislature's investigative commission turns up criminal behavior on his part, it can send Samper to face the Supreme Court.

The creation of this commission has the advantage of calming people down. Colombians tell themselves that they're finally going to learn whether or not their president is responsible for having financed his campaign with drug money. But I'm convinced that the commission has been created mainly to exonerate Samper—to forestall, at whatever price, the terrible catastrophe that is threatening the whole political class.

The commission hasn't even begun its work before a second wildfire breaks out. Over the weekend of July 29–30, Samper's two closest collaborators—his minister of defense, Fernando Botero, and his minister of the interior, Horacio Serpa—go to the president's secondary residence to examine Valdivieso's case against Medina, the imprisoned treasurer. On Monday, July 31, Botero and Serpa call a press conference to explain that Medina's allegations don't hold up, and that if he's accusing the president, it's only in the hope of receiving a lighter sentence.

What happens next is a dramatic reversal.

"You say that Medina has admitted this and that," a journalist says in surprise. "But how do you know, since the investigative file is sealed?"

This is an unprecedented television moment in Colombia.

"We've gotten this information," Botero begins, "and the minister of the interior is going to tell you how."

Then, brusquely, he hands the microphone to Serpa, whose face turns purple. Naturally, they can't admit that they've stolen the investigative file over the weekend.

Serpa hesitates, stutters.

"An anonymous source brought it to the interior ministry," he mumbles sheepishly.

The lie is so obvious that the next day, in the name of the independence of the judicial system, columnists are demanding the resignation of these incompetents. Thus a secondary scandal takes shape (two ministers' violation of the secrecy of a judicial inquest) within the main scandal (the narco-financing of the president's campaign).

On August 2, the star minister, Fernando Botero, resigns.

Once again, it's a matter of finding a way to calm people down as soon as possible. For me, it's clear that Botero, acting as a faithful lightning rod, is taking the hit in order to save Samper. Whatever the cost, the president's name cannot be brought into this ugly, pitiful affair.

On August 4, the legislature's investigative commission officially begins its work. The prosecutor Valdivieso has sent it all the evidence that will allow it to form its own opinion regarding the degree of the president's involvement in the enigma of his campaign financing.

The commission however, has hardly opened its sessions when a first, explosive response to its questions is given on television: on August 5, on a weekend, the press recalls the existence of a very damaging recording involving the narcos. It is a conversation between Samper and a certain Elisabeth Montoya, a member of the drug mafia, in which she announces to Samper that emissaries are coming to see him. The magazine *Semana*, which publishes the whole conversation, makes the announcement of its scoop. The tone leaves no doubt regarding the friendly ties between Samper and this woman—on several occasions, she calls him by his nickname "Ernestico"—but above all it's clear that emissaries from the mafia are bringing money to a man who is still only a candidate for the presidency.

The publication of this document, following the admissions made by the treasurer, Medina, is catastrophic for Samper. It comes at a time when with Botero's resignation, the highest levels of government seem to be stumbling, losing their grip.

What will Samper say? What can he come up with now to escape this cataclysmic whirlwind spinning out of control?

The next day, on Sunday, August 6, Miguel Rodriguez, the high-strung Miguel, is captured. I have in my mind the image of Miguel listening to his brother Gilberto talking about his surrender, the years in prison, and shouting, "Well, I'm never going to jail. I'd rather die." Obviously, he has surrendered because only the spectacular announcement of his capture is still capable of saving the president, of causing a turnaround in public opinion. Has Samper been heard bantering with a mafia woman? Sure, but that won't keep him from throwing the bosses of the mafia in prison! But why should the Rodriguez brothers help him out? Did they make with Samper the agreement they didn't have time to conclude with Gaviria, his predecessor: a surrender without extradition, a few years in prison to make it look good, and then wealth and happiness for the whole clan? According to sworn testimony by Fernando Botero and Santiago Medina in trial 8000, Samper guaranteed the Rodriguez brothers safety in exchange for payments made to his campaign. Samper is the saviour to the Rodriguez brothers. If he goes down, they'll follow him there and find themselves under fire again, threatened with real imprisonment for life and, above all, with the probable return of extradition. That's why they continue carrying this moribund president on their shoulders by sacrificing themselves, one after another, hoping to get out again as soon as Samper is back on his feet.

The purported arrest of Miguel Rodriguez produces the desired effect in the media: it occupies the front pages of the newspapers and relegates to the inside pages the commentaries on the telephone conversation between the president and Elisabeth Montoya.

Ernesto Samper can proudly claim to have decapitated the Cali cartel. Are the people duped? They seem to be living with their ears glued to the radio, groggy, expecting a new part of the sky to fall on their heads every morning. What can they do to take their destiny back into their own hands? Nothing. Against their will, they've set out on a ship sailing directly, madly, into dense fog.

On August 15, another thunderbolt: the ex-minister Fernando Botero is indicted by Alfonso Valdivieso for false testimony and illegal enrichment, and is thrown in prison. By resigning thirteen days earlier, he has lost his ministerial immunity. The fall of the man who, the year before, had incarnated the Galil scandal seems to be a strike of justice. But in view of the scope of the crisis the Colombian government is undergoing, Botero is no more than one pawn among others.

I think of his father, the painter Fernando Botero—a friend of my parents, extraordinarily kind and selfless throughout his life, known to everyone for his integrity—and find myself feeling a deep sorrow for him. He's a man I love and admire. It's not fair; he doesn't deserve this, not as a father, not as a famous Colombian who has given pride to our country.

A pawn, I said, but one who is nonetheless treated with immense respect. Botero is not imprisoned with thieves and criminals; for him, the riding school has been requisitioned, and on television we see him riding through the grounds, playing with his children. Like him, all the legislators imprisoned in the course of trial 8000 will enjoy privileged treatment. Not to mention the Rodriguez brothers, who continue to live in insolent luxury, while other detainees experience a daily nightmare in the overcrowded prisons, where they must go so far as to pay for the "privilege" of sleeping lying down. What can Colombians make of this? It's clear that the system is not constituted in such a way as to dissuade this indecent marriage between the political class and the mafia. Even in prison, the mafiosi get their due.

The next day, August 16, a tragedy chills my heart: the driver of the interior minister, Horacio Serpa, is assassinated on the street. Not just anywhere, either, but within twenty yards of Valdivieso's offices, where he was headed. For me, this assassination is a terrifying revelation: this man, Dario Reyes, is killed at the moment when he was going to testify. Later, Manuel Vicente Pena wrote an article for *La Prensa,* followed by a book titled *El Narco Fiscal* ("The Narco General Attorney") in which he contends that Serpa, the minister of the Interior, is closely linked to this murder. When the article comes out, it rings an alarm bell in my head. I am now convinced that these people will do anything to maintain their power.

Bogotá, February 1962.
In the arms of my
mother, Yolanda Pulecio.
I am three months old.

Bogotá, 1963.
With my mother and
my sister, Astrid.

Bogotá, 1967. My mother becomes the assistant for social affairs to the mayor of Bogotá, Virgilio Barco. Here with the president of Colombia, Carlos Lleras (left), and Barco (right).

Bogotá, 1985. A niño sleeping in the street.

The Seychelles Islands,
November 1985.
Melanie with her
father, Fabrice.

The Seychelles Islands, December 1985. With my mother and father.

Los Angeles, December
1988. With baby Lorenzo.

1994. Campaign headquarters for my candidacy for the House of Representatives.

**LEFT** Bogatá, March 1994. With my friend, Clara Rojas, celebrating my election to the House of Representatives.

**BOTTOM** Colombia, 1994. With my "Musketeer" friends Maria Paulina Espinosa and Guillermo Martinez-Guerra.

**ABOVE** Bogotá,
April 1994.
The official family
portrait after the
House elections, with
Melanie and Lorenzo.

**ABOVE** Bogotá, 1994.
With grandmother
Nina, Melanie,
Lorenzo, and the
little rabbit.

**LEFT** New Zealand,
January 1997.
With Melanie and
Lorenzo in Auckland.

**ABOVE** Colombia, March 1998. Campaigning for the Senate.

**TOP** Auckland, January 1997. Melanie and Lorenzo in their new school uniforms before entering Churchill Park School.

**MIDDLE** 1997. With my husband, Juan Carlos.

RIGHT March 1998. Official photograph for the election ballot.

BELOW March 1998. Celebrating my victory in the senatorial election, when I was elected senator with the highest score in the country.

May 6, 1998. With the presidential candidate Andrés Pastrana, signing the anticorruption pact for a major political reform through referendum.

Bogotá, November 2000. Marching against electoral fraud.

2000. With my parents.

# CHAPTER FIFTEEN

**AS SOON AS** I hear about the assassination of Horacio Serpa's driver, I decide to break all ties with Ernesto Samper. I will no longer go to the legislators' meetings organized by the presidential palace. I will decline all invitations to hold private conversations. It has become absolutely necessary for the judicial system to make a statement regarding this man's responsibility, and my job, as an elected representative, is to help it do so.*

At this point, I feel I have to play an active role in what is happening to us, Colombians. I know that I have to denounce systematically, whenever I have proof, the lies told by those in power. I must expose what they conceal. Further, I feel I have to assume the role of saying publicly what Colombians—who are stunned, humiliated, and reduced to silence—think in their hearts. More than ever, I feel committed to keeping a daily record of the acts and deeds of the men who govern us. I'm preparing myself to enter a very dangerous battlefield.

---

*As members of Parliament, we become the president's legal prosecutors. Our role is to scrutinize, to evaluate the case against the president. We cannot absolve or condemn him, but we can evaluate the evidence against him (testimonies, documents, etc.) to see if they are strong enough to warrant an indictment. If it is the case, the president goes to trial before the Supreme Court.

Events unfold at a dizzying speed. On August 31, 1995, two weeks after the "imprisonment" of ex-minister Botero, the Liberal Party makes a decision, at Samper's request, that represents the height of cynicism: in the name of the code of ethics I myself wrote, it suspends Fernando Botero and the ex-treasurer, Santiago Medina. That's how Samper is trying to shore up his latest version of the facts, according to which his two collaborators are supposed to have accepted money from the mafia "behind his back."

On September 4, however, this pitiful maneuver is pulverized when several newspapers publish the first irrefutable evidence against Samper: a facsimile of a check for thirty-two million pesos signed by . . . Elisabeth Montoya, the woman who, a month earlier was heard on tape addressing Samper by the affectionate diminutive "Ernestico." This in itself should have been sufficient for the same sanction to be imposed against Samper.

But will Colombians take to the streets to demand the departure of a president whose disgrace reflects poorly on the whole country? No, they remain quiet, as if petrified by the horror of it all. A year earlier, the revelation of the cassette on which we heard the Rodriguez brothers praising Samper had aroused indignation; people wanted to believe in their president's integrity. This time, confronted by the mounting evidence against him, they no longer know how to react, or whom to trust. They seem floored, resigned to the worst, like those martyred people whose dead are totted up and whose voices of protest are silenced. I feel sorrow for those who are being trampled, an immense revolt is brewing inside me.

The fallout goes on and on: on September 19, we learn that the Cali cartel's accountant, Guillermo Pallomari, has just fled Colombia to seek the protection of the United States. If this man, who knows everything about the financing of Samper's campaign, has left, it's obviously because he's received death threats. Someone feared that he might decide to testify in court. And for those who are still skeptical, Guillermo Pallomari's wife, who remains in Colombia, is killed a few days later. The message sent to the accountant is this: if you talk, we'll kill your whole family. Nonetheless, this man who is now sheltered by the gringos represents the greatest current danger to Samper, all the more so because Alfonso Valdivieso is officially requesting from the United States an authorization to question Pallomari. The press is already announcing that the investigation will be pursued in the United States.

What can Samper come up with now to reverse the tide of history? I think about nothing else, and, a week later, I get my answer. On September 27, an assassination attempt is made on the president's attorney, *el doctor* Cancino. Samper had named this attorney to represent him before the legislature's investigative commission. Oddly, Cancino escapes with only a slight scratch on his finger, whereas his bodyguards are all killed, their bodies, riddled with bullet holes, testifying to the ferocity of the shooting. We are told that Cancino, pursued by the killers, managed to get away.

But this version doesn't convince me. Cancino is a paunchy, elderly gentleman; I find it hard to imagine him running up Bogotá's steep streets to escape young *sicarios*. This is a setup; what's behind it? Horacio Serpa, Samper's most faithful accomplice, provides the answer the following day. During a televised press conference on the circumstances of the attack, the interior minister is asked: "Do you think the United States was involved in this incident in some way?" The interior minister ponders the question for a moment and then says: "It's possible, it's possible; in fact, we have to give serious consideration to that possibility."

What a good move! That's all it takes for people to start saying that everything is happening because the Americans are plotting against Samper.

The next day, the press excitedly inflames this national sentiment, fanning the flames of xenophobia many Colombians feel with regard to "gringos." And it works! Colombians, who are deeply disappointed and terribly silent, ask nothing better than to believe these pitiful concoctions. They regain hope, they straighten their backs, and while the highest levels of government are collapsing, the county is engaged in one of the most violent episodes of xenophobia that has occurred in Colombia in recent years. Yes, the United States wants to bring down our president, they want to destabilize our democracy, they cannot bear our independence, they want Colombia to be servile, colonized . . . During these delirious, dreadful weeks, foreigners, whoever they might be, are having a hard time. In the newspapers and on the radio, the only thing we hear and read is how Colombia's independence is being threatened by the American eagle. Yet again, innocent people are suspiciously killed by assailants who will never be brought to justice.

On November 2, a new attack throws cold water on this hateful and xenophobic attitude: the conservative leader Álvaro Gómez is assassinated. For the past year, Gómez has been the only major politician openly demanding Ernesto Samper's resignation. The whole country knows about Gómez's close ties to the United States—so close that he once seemed to be Washington's candidate to succeed Samper. The son of Laureano Gómez, the infamous Colombian "Duce" (president from 1950 to 1953), Álvaro Gómez nevertheless became a prominent figure in Colombia. He was a presidential candidate several times and served as ambassador to France. At the time of his death at the age of seventy-five, he enjoyed an intellectual and moral aura that placed him, in the minds of Colombians, above the traditional political class. When consulted, he did not hesitate to tell the most disturbing truth straight on. To many people, he was a final recourse.

His assassination throws people from revanchist frenzy into terror. It signals that something has gone mad in Colombian society, that certain people no longer respect any principle, and will do anything, including murder, to accomplish a goal. Killing Gómez is a declaration of war on the United States and, within our borders, on anyone who dares to challenge the status quo. From now on, any threat to the regime will mean death.

Through all this, the allegations against the president still remain central in the public's mind, and the people begin to wonder whether the imprisoned Fernando Botero, the one individual who would intimately know the details of any drug financing, would reveal secrets. The former defense minister is dangerous because he's still powerful, and because he's in prison and might be tempted to spill his guts in order to get out. Up to this point, Botero has kept quiet, unlike the former treasurer. To me it seems clear that he's waiting for Samper to make an overture to him.

Sure enough, on December 10, we learn that the head of state has made a lengthy visit to his former comrade. In other countries, it would surely be unexpected for a sitting president to go to a prison to talk with a former member of his government suspected of having illicitly enriched himself through his connections with the drug dealers. But it's no surprise to Colombians. Nothing surprises Colombians any more. Besides, the television cameras are there, and they show us the two men spending four hours strolling gravely up and down the walkways in the garden of the riding school where Botero is being held.

What are they talking about? The answer to this question comes a few days later, but it comes in so many wrappings that it takes the perspicacity of a man like Valdivieso to sort it out. Back in the legislature, while voting on a bill, the Senate adopts a tiny article that has nothing to do with the text of the bill but which stipulates that henceforth no one can be prosecuted for illegal enrichment unless the illegal origin of the funds has been proved. Since the Rodriguez brothers have not yet been tried, there is still no evidence that the money Botero received is illegal and came from them. Thus, if this article is approved by the legislators, it will have the force of law, and Fernando Botero will be immediately set free.

So that's what they have been preparing in the gardens of the riding school. This is what Samper has come up with to save his accomplice and dissuade him from talking. Colombian politicians are accustomed to adding articles to legislative bills whose only object is to serve some individual's interests. They are called *micos*, "monkeys," because they can conveniently hang like monkeys from any branch of the legislative text and often go unnoticed. This one will be nicknamed *narcomico*, but it won't go unnoticed. On the contrary—it will give me an opportunity to confront the regime head-on for the first time.

The *narcomico* comes before the representatives on December 15, only five days after Samper's visit to Botero. What strikes me right away is that the legislative chamber, which is usually three-quarters empty, is completely full that afternoon. Then I recall Gilberto Rodriguez's proud admission: "Most of your fellow representatives are in our pay, *Doctora*. Would you like their names?" It's pointless. The extreme tension in all these men shows that they've gotten the message, that they're controlled by the same command center higher up and, ultimately, by the same interests.

I observe this circus with deep revulsion. Before starting the proceedings, the representatives are waiting for the minister of justice, Humberto Martinez, to appear. I'm also waiting for him, and—the better to welcome him—I've even posted myself near the door through which he'll enter.

"Bravo for the *narcomico*," I say to him. "You're negotiating Botero's release, and you can count on me to make that known."

I can tell that Martinez is completely disconcerted by my words. But he's a clever man, and he knows how to duck.

"I'm just as concerned as you are," he murmurs.

"Then say so! You didn't speak in the Senate."

"The government gave orders but they gave me no support. Serpa didn't come to help me. You speak, Ingrid, you can help me."

"You bet I'm going to speak."

At that precise moment the doors open, and who do we see coming in? Serpa, the interior minister, the brilliant, zealous spokesman for Ernesto Samper, the very man who slipped up during the preparation of the case against Medina.

I rush up to him. Mockingly, I say:

"Just the person I wanted to see! I hope you've taken the trouble to get that indecent article removed."

He seems breathless, mumbles a few inaudible words like a man who knows he's been caught red-handed, and moves away.

The debate begins.

Humberto Martinez speaks in favor of eliminating the *narcomico*, but weakly, as if this whole affair is just a small detail in the text on which we are about to vote. He makes his lack of enthusiasm appear as prudence or even respect for the House. However, his nervousness and uneasiness send ripples of panic among the representatives. The tension goes up several degrees, and I can see all these corrupt men anxiously counting their own votes. They've nothing to fear; clearly, they're in the majority. But this tension, which can also be discerned in Serpa, intrigues me. I sense that something I haven't been aware of is going on behind the scenes. I go to find out. I leave the assembly and talk with the journalists, who are crowded behind the door, waiting to take a peek at what's going on inside. And there's a dramatic development: I learn that the legislature's investigative commission, which is said to have been looking since August into Samper's dossier, is about to render its verdict! That's why Horacio Serpa is there.

I am expecting the worst. I know the commission is there to save Samper, no matter what. My absolute priority is to vocally denounce, as soon as possible, the deal with Botero behind the narcomico, before they exonerate Samper and close the investigation against him for good. I want the Colombians to know the truth, and I now have the information to reveal the plot.

I'm given the floor.

"Do you know why our president went to talk to Botero in his cushy prison?"

People's faces freeze over. The press is there, and they can no longer shut me up. So I reveal the secret pact Samper and Botero have made. Then I shout:

"We all know that if the *narcomico* is adopted, that will be the end of trial 8000, and the end of problems for most of you and for those politicians, your friends who are already in jail. We all know that many in this house have received dirty money, just like Botero. Colombians have the right to know that if their representatives are all here today, it's not to defend our laws, but to avoid their enforcement, to save their own skins, to escape punishment. Why is Minister Serpa silent? Why doesn't he come up here and tell us clearly what is the government's position on this iniquitous, disgraceful article? Because the government has given instructions to you all to vote on this article."

An unprecedented outcry erupts in the assembly. A year earlier, I was seen crossing swords with Botero on the Galil rifle scandal, but back then I was not challenging the legitimacy of the government. This time I am. Not only does the scandal tarnish the government and the legislature, but here I am, accusing the president of the republic of being chiefly responsible for it.

I assume that Serpa is going to jump to his feet and mount the podium to attack me as soon as I've said my last word. He's a formidable orator, a man to whom people listen with pleasure because he can be flamboyant—and lethal. But he doesn't move. What's he waiting for?

Other speakers follow me. Almost two hours go by. Then the doors of the assembly suddenly burst open under the pressure of a crowd of excited reporters. They're looking for Serpa, who immediately understands this and joins them. Representatives get up and run outside, following them. I also hurry over.

The investigative commission has just made its decision public: it refuses to accuse or absolve Samper, on the pretext that there's no evidence for them to decide whether he is guilty or not. In other words, it disqualifies itself. They are awfully good, I say to myself. They found the legal way to avoid impeachment. So Samper is saved, he won't be investigated. The commission has met only in order to suppress the scandal. When he hears the news, Serpa nods and says with marvelous hypocrisy:

"We respect the legislators' opinion. Calmly, taking all the time they needed, they've studied the prosecutor's report and they've concluded that there is no charge to be made against President Samper. I'm pleased. As for the rest, I put my trust in the Colombian judicial system."

As soon as he returns to the floor, Serpa mounts the podium. And in the name of the independence of the judicial system, he demands the withdrawal of the *narcomico*, which was introduced in the Senate, without, he claims, the government having been informed. In a second, I understand that since he is now master of his fate, Samper is abandoning Botero. Appealing to great principles, Serpa goes on:

"And we do not accept the moral lessons proffered by that impassioned legislator, Ingrid Betancourt, who dares to spread malicious gossip and suggest that the head of state and the government are guilty of dishonorable behavior."

Seeing red, I demand the floor to reply. Serpa pretends not to have heard me. I raise my voice, but he goes on, imperturbably. I stand up, and Serpa continues, but no one is listening to him any more: all eyes are on me, especially those of the journalists. Our internal rules guarantee my right to reply. Serpa is in the wrong by refusing to let me speak. Since I won't give up, he finally yields, furious:

"All right, let's hear your reply, please."

"What the country wants to know," I say, seizing his microphone, "is not what you think of me, but why you didn't tell the Senate ten days ago what you're telling us now, today, in the House of Representatives. What the country wants to know is why the government's position on the *narcomico* changed the minute you took notice of the legislature's investigative commission and the announcement of its decision, which saves only the skin of the president of the republic."

And I throw the microphone in his face, so that he jumps backward to avoid getting hit by it.

The press has understood, and so have my colleagues. Now that Samper is safe, the government is abandoning them, and many of them are going to end up in prison. Botero has just lost his last battle. The government is letting him go down alone. Serpa is well aware of the danger; the silence in the assembly is threatening. He decides to adopt a completely different tone:

"Pardon me if I've offended you; perhaps I didn't express myself well. If you're asking for a clarification, I'll be glad to give you one." And so on.

What he says doesn't matter anymore. My feeling is that I've scored a point by publicly unmasking them.


We're on the eve of the long Christmas vacation. The legislative session is coming to a close, and it's with relief that I look forward to the coming weeks of rest, especially because I'm sure that Botero's going to talk. I know his ambition, his high opinion of himself—he was dreaming of being president of the republic—and I don't for a moment believe that he's going to let himself be sacrificed and bear the whole weight of the scandal simply to save Samper.

The solidarity of the government itself is crumbling: after hypocritically congratulating me, the minister of justice, Humberto Martinez, whom I had sensed to be ill at ease, resigns. As a reward for his silence, he is made ambassador to France.

# CHAPTER SIXTEEN

JUAN CARLOS, whom I'd met a few months earlier and who would become my future husband, and I are going on our first vacation together. We want to have a little time to ourselves. For a few days, we stay in Bogotá. The whole city has been deserted. Everyone is away on holiday, and we are enjoying the city at its best. Life seems easier, more lighthearted. Then we leave with the children for Tairona Park. It is a beautiful spot on our Carribbean coast, right by the Sierra Nevada, the world's highest mountain on a waterfront. This is a magic place, and we feel like we are on another planet. We stay in one of the thatched cabins on the shores of the Caribbean with the virgin forest behind us. There is no radio, no television; war, violence, the drug traffickers, and Samper all seem very far away. Yet the few tourists who recognize me want to initiate a political conversation. I am forced to hide. I need to recover my role as a mother, fully, exclusively. I have to run away, to isolate myself. Juan Carlos supports me tenderly. We make sand castles, collect seashells, and take the children into Tairona's enormous waves. Born in Cartagena, Juan Carlos is the best surfing teacher. He grew up riding the waves.

In the evening, without electricity, we look at the stars. This is the time to tell stories; Melanie wants to hear about the first months of her life, what she was like as a baby, what her first words were. Lorenzo wants stories about bears and foxes, the ones his father made up for him. They fall asleep, happy and confident.

We come back to Bogotá rested and calm. One evening in January, when Juan Carlos and I are still on vacation and alone, we decide to go out. He has a vague desire to go to a movie, and I'd like to get some fresh air. As soon as we go outside, we're struck by the strange silence that reigns in Bogotá; you'd think you're in a ghost town. We drive down almost deserted avenues that are normally jammed with rattling buses, trucks, cars. Suddenly, just as the riding school where Botero is "imprisoned" comes into view, we're astounded to see that assault tanks have taken up positions around the walls of the compound.

"Juan Carlos, look!"

"Oh, shit! What's going on?"

"I don't know. Something serious, that's for sure."

"It looks like a coup d'état, something like that."

"Of course! That's why there isn't anyone on the streets! We are the only fools out here. Let's go home quickly, and listen to the news."

It's Tuesday, January 23, 1996, a date that will go down in the annals of Colombian history. On television, Fernando Botero, interviewed by Yamid Amat, is lambasting President Samper. Yes, the Cali cartel financed his campaign; yes, Ernesto Samper knew all about it.

"And what about you?" Yamid Amat asks.

"I was not brought in on this. I didn't know anything about it."

Juan Carlos and I look at each other with consternation. Who could believe that Botero didn't know what was going on? But it doesn't matter; the important thing is that this time the accusation is coming from inside the president's circle, and Samper himself has no way of shutting up Botero. By bringing tanks up around his "prison," the former minister of defense has shown that he has the support of the army.

Wounded, Samper reacts the next day by calling his former accomplice a traitor and a liar. He reiterates that he didn't know anything, that the transaction took place behind his back. He asks his former defense minister to answer this apparently reasonable question: How can Botero accuse him of knowing all about a crime Botero himself claims to have known nothing about?

Samper is fast. This is clever, to be sure, but it's not quite enough to convince a mistrustful people or the international observers whose confidence in the president is steadily declining. What can Samper do to get control of the situation? Since he's not resigning, he must be working on still another attempt to win over the most dubious.

On January 30, Samper's plan is already in place, and he gives the whole country a jolt by calling for a special session of the legislature to determine whether or not he's guilty. In other words, he foregoes the friendly neutrality of the legislature's investigative commission and demands that his case be heard in plenary session. "I have been acquitted," he says, "but after the new accusations that have been made against me, doubt has crept into people's minds. I owe them the truth." And, to win the support of those who might still doubt his good faith, he goes even further by asking that the whole trial be covered live on television. This is a first; Colombian television has never been authorized to film a full session of the legislature. And what a session it would be: the accusation of a sitting president!

In the name of disclosure, Samper demands that the representatives judge him. But Samper is not running much of a risk if, as the evidence suggests, many of these same representatives are likewise tainted by the drug financing of their campaigns. Obviously, it's no accident that he has specifically requested that the representatives decide his fate by a show of hands. On the pretext of preserving people's right to know, he's making sure that no one will betray him by taking advantage of the secrecy of the ballot box.

How can this be exposed? How can Colombians be made to understand that once again they're going to be fooled, that their ignorance of the political world and of its extreme venality is being taken advantage of? Certainly, the plenary session will debate the president's responsibility, but the debate will be introduced and controlled by the same representatives in the investigative commission that has already so obligingly declared itself incompetent to pass judgment on this situation. So it's this commission that has to be recomposed, but how?

I'm one of several independents who feel discontent at what's about to happen. We decide to get together a few days before the opening of the plenary session. Among us—ten to begin with, though our numbers will soon dwindle—are Maria Paulina Espinosa and Guillermo Martinez Guerra, two of the "musketeers" from the Galil affair. Also with us is Viviane Morales, an energetic woman who proposes that we begin a hunger strike to demand that some of us be put on the investigative commission.

"I agree," I say. "Let's make a solemn vow to pursue a hunger strike inside the chamber until new members have been named to the commission."

Empowered by this commitment, I take the floor in the plenary session:

"For the future of Colombia, it is essential that the trial which is opening here be absolutely fair, for the implications go far beyond the fate of President Samper. What is at stake here is our right to know the truth, our right to write our own history; for the members of the House, what is at stake is whether we will be able to look at ourselves in the mirror tomorrow morning without being ashamed. The commission, as we all know, is composed of the president's most faithful friends. Several of us believe that a fair trial thus requires the appointment of a new commission. For us, this is an indispensable preliminary.

I tell you here and now: we're prepared to go on a hunger strike if the assembly refuses to name new members to the commission."

Outcry, insults, unprecedented chaos follow. The session is immediately closed, and when silence falls on the chamber, only two of the ten who committed themselves to the hunger strike remain: Guillermo Martinez Guerra and myself. We soon discover that Viviane Morales, the enthusiastic initiator of the strike, was received by Samper the day after we met and changed her mind, seemingly in exchange for secured government jobs for members of her family.

So here we are on a hunger strike. How long will it last? Two or three days, I tell myself; they'll give in, it's impossible otherwise. We have to get organized to hold out. We both call our homes to have toiletries and blankets brought to us. Then we mark off a space in the chamber and set up our camp.

The first night is highlighted by moments of euphoria. We're confident. Guillermo Martinez Guerra, a former fighter pilot, has nerves of steel. But the next day, as the legislative chamber fills and work resumes as if we didn't exist, we begin to feel nervous. We have the feeling we're tilting at windmills. From the knowing, quietly ironic smiles of our colleagues, we deduce that they think we won't hold out for long. Moreover, our hunger strike is not reported in the press.

On the third day, the press gets involved after all, but, as one might expect, in the worst possible way. Television shows the delivery of trays of chicken, claiming they're intended for us. The commentators make fun of us: we're on a hunger strike during the day, they claim, but at night we stuff ourselves. Or they say something like: while people are really dying of hunger, an upper-class woman is engaging in blackmail by refusing to eat in order to get appointed to a commission.

The scandal escalates. Whether they think we're sincere or not, Colombians can no longer ignore the fact that two representatives, two of the four "musketeers," are officially on a hunger strike for reasons that the journalists are doing their best to ridicule. Policemen, sent by the minister of the interior, keep an eye on us day and night. That, at least, shows that concern is slowly spreading through the government. They know that our strike is not a fake, and that a day will come when it will no longer be possible to distort the public opinion. Time is on our side.

I encounter my first real test on the fifth day: my seven-year-old son, Lorenzo, refuses to eat and vomits the little that he takes in. He's quickly becoming dehydrated, and the doctor sees no solution other than to hospitalize him. On the telephone, my mother sounds horrified, her voice broken by sobs.

"Bring him to me, Mama, please, as soon as you can. I have to talk to him. I have to explain to him that I'm not letting myself die, that on the contrary, I'm fighting; otherwise he'll stop eating in order to imitate me."

It is true that I'm fighting with all my strength. And I need Lorenzo to witness this. He finds me literally besieged by the radio and television journalists who've been dragging us through the mud since the first day of the strike. Revolted, beside myself, my heart broken by the harm all this is doing my son, I'm explaining to them that I wouldn't be there if the press did their job, if they denounced the corruption of this government instead of covering up its mistakes. The climate is feverish, indescribably confrontational. The police are watching, and the journalists, so full of themselves when they're alone with me, suddenly seem paralyzed. My boy is there, fixing the journalists with a dark stare. They step aside to let Loli come toward me.

Has he understood that I'm on the side of life? That his mother does not want to die, as people have been telling him at school?

"Why am I not eating, Loli? Because I want people to listen to me. Guillermo and I know that very bad things are being done in the government, but people won't listen to us. Sometimes you throw yourself down on the ground and scream when I don't listen to you, right? Well, today, I'm doing more or less the same thing. At my age, it doesn't matter whether you eat or not, and it's hard for me just because I like to eat, but I'm not going to die, Loli. On the contrary, I'm going to win, and soon Colombians will listen to me and believe what I say. But to go on I have to be strong and courageous, and I won't have the courage I need if you don't eat. Listen to me, Loli: I need you to eat so that I can go on. You understand, I need you to eat. Help me!"

"Then you're not sick?"

"No, of course I'm not sick! I'm fine. This is a decision I've made because I'm angry. When people start listening to me, I'll stop being angry and I'll come home."

For two hours, we talk in this way, and Loli leaves me reassured. He has understood my battle—my battle is proof that I'm alive. That very evening, he eats his dinner as usual. He doesn't have to go to the hospital.

For me this is a first, immense victory.

The second victory is attained by the slow turnaround in the press's attitude toward us. Our hunger strike has elicited growing concern abroad, and we have the feeling that Colombian journalists are discovering its importance by reading the international press. So—French, German, American, Japanese journalists who are coming to see us are not laughing? Maybe we'd better have a closer look, after all. And these people who had openly mocked us move from incredulity toward a neutrality that is first benevolent and then admiring. They decide to acknowledge that the battle I'm waging has a certain coherence, from the condoms against corruption in my election campaign to this hunger strike, by way of the revelation of the Galil scandal.

Finally they recognize me, they understand our commitment. To the two of us shut up inside this chamber, it means that for sure Colombians will finally understand what's going on and will support us. It also means we're scoring points against our "colleagues" in the House of Representatives, whose stubborn refusal to change the commission now seems increasingly suspect.

It's to my fellow legislators that I address myself, in plenary session, after a week on hunger strike. The assembly is still debating matters of procedure. Over the preceding days, while our struggle is not taken seriously by the press, our colleagues, or the Colombians who watch the disinformation on television, Samper has obviously invited to the presidential palace everyone who has any power in Colombia—big industrialists, political leaders, labor unionists—and they have all assured him of their support, because none of them dares to bank on his fall, and they all believe Samper will know how to be grateful to them. The representatives are also kept in line by an appetite for gain. That's what I want to tell them, looking straight in their eyes, hoping to reach the Colombian people beyond them.

"You're taking advantage of the people's hopes. This is not a trial we're embarking upon, it's a comedy, a pathetic and useless comedy, and you're well aware of that. You're pretending to judge a man, but you all have a personal interest in letting him off unscathed. Tomorrow, he'll throw each you a bone to thank you for your support."

Never have I spoken to them so bluntly, and yet never have I felt more serene, more sure of myself.

Fasting has given me a feeling of unreality, and also a feeling of control, as if I weren't really here among them, as if I were drawing my strength from a higher, inexhaustible source. And these people who haven't spoken to me for months, even to say hello, who hate me, are touched in spite of themselves. It's as if they were put under a spell. Not one of them interrupts me. I speak, and I say what can't be said. A few hours later, in the corridors, ashamed by their own silence, many of them accuse me of "moral blackmail." Those who are most angry with themselves even tell journalists: "She can just go ahead and die, we won't give in."

This is not true. Actually, they are beginning to lose their grip. The negotiations to find a way to make Colombians believe that the commission is not composed solely of Samper puppets have already begun. Three politicians from the opposition party are put on the commission, but they have been chosen carefully, so as to be suffi-ciently corrupt not to pose any threat. The press noisily celebrates this move and says that we have succeeded in changing the commission. What do I think about it? "This is just another trick, and you know it perfectly well, but you're careful not to say it in print."

It is in this context that I receive a visit from my father. He's just come back from a long trip abroad and hurries to the legislature. While from the outset I've seen in Mama's eyes nothing but incom-prehension and concern, in the serenity of Papa's eyes, I see pride. Oh, the way he looks at me! The moment I run into his arms, my heart swells with joy. Papa sits down, takes my hand. "Now that you've gone this far," he says gravely, "there are only two solutions left for you: either you win and you go out of the House holding your head high, or they don't give in and you have to go all the way, Ingrid, all the way. You have to prepare yourself for that." Yes, my father's solidarity with me, his support, goes so far as to envisage my death. No declaration could strengthen my will more and surely liber-ate me by showing me more clearly the path I have to follow.

# CHAPTER SEVENTEEN

**IS IT AN ACCIDENT** or a cleverly thought-out strategy? Repairs to the legislative chamber begin during the long hours when the House is not in session. The ceiling is redone and the window frames are removed, so that at night an icy wind whistles around us. They make plaster and cement, and drafts blow around an acrid, unbreathable dust. Weakened by ten days of fasting, I fall ill. My pulse grows weaker, I can't get enough oxygen. The doctors treating me decide that it's dangerous for me to remain in this air. I refuse hospitalization, and oxygen tanks are brought. I'm now surviving by lying prone under a mask that I remove when journalists pass by.

Finally, what the doctors feared occurs: after two weeks without food, prostrated by fever, I fall into a state of deep unconsciousness from which I emerge only in the hospital, with an IV in my arm. My father is there, and our eyes meet again, at the very moment when I realize what's happened. Tears come into my eyes; I'm so sorry not to have held out to the end. And then he says:

"You went as far as you could, Ingrid, until your body betrayed you. You've done well, you can be proud. In any case, I'm proud enough for both of us."

He smiles, and I try to smile back at him. My feeling of failure is so intense that I'm overwhelmed by sobs.

At this very moment, I'm convinced that all this suffering—mine, but also Loli's—was in vain, that all these sacrifices we imposed on ourselves did no good. But I'm mistaken, as I discover over the coming months. My hunger strike opened the eyes of the Colombian people. They got my message, understood that they were being tricked, that the game was rigged from the start. Without my being aware of it, a relationship of confidence has grown between them and me during these two weeks. They will never forget how atrociously the press behaved, how much it lied, especially by claiming that I was secretly eating. From now on—I will have many occasions to observe this—the journalists can say whatever they like about me, but the Colombian people will believe only what I tell them. They can find twenty people to contradict me, they can organize television panels and put me on trial, but nothing will work.

In this regard, the hunger strike marks a crucial turning point in my life: it seals a special bond between Colombians and me, a bond that will be proof against any future campaign to defame me, and that will allow me, two years later, to be elected senator with more votes than any other candidate, and without the support of any political party.

While the representatives are getting ready to proceed with the rigged trial of the president of the republic, the judicial system is pursuing its work under the aegis of the prosecutor, Valdivieso. The judicial system does not have the power to indict and judge the president, who enjoys immunity from prosecution, but it has the right to take the testimony of all the witnesses and thus implicitly establish his guilt. The treasurer, Medina, and the former minister, Botero, have talked, and the damage is catastrophic for Samper. He doesn't want any more witnesses to be heard. Horacio Serpa's driver has already paid with his life for trying to testify.

On February 1, 1996, Elisabeth Montoya is assassinated. This woman, who held all the evidence of Samper's guilt, and a copy of whose check paying him thirty-two million pesos the press had already published, is found dead in an apartment in the southern part of Bogotá, killed by two bullets fired into her vagina. An attempt is made to make us believe this was a crime of passion. In reality, as I soon confirm, Elisabeth Montoya, abandoned by the Rodriguez brothers and harassed by Samper's thugs, had contacted Valdivieso. She was living in fear, and she no longer knew how to escape the death sentence she felt was coming. During the weeks before her death, she'd tried to collect all the banking records proving the transfers of funds to accounts controlled by Ernesto Samper. She hoped to flee Colombia with these documents, which, she was convinced, would provide the best protection for her.

On March 1, a few weeks before the opening of the chief of state's "trial," the U.S. government sends the country, and the world as a whole, an unequivocal message: they "decertify" Colombia. In other words, they declare Colombia out of bounds for further financial aid.* There could be no better way to show to show us, Colombians, that the United States has arrived at the conviction—and probably have the proof to support it—that Samper was in fact financed by drug money.

In a democracy worthy of its name, this kind of international repudiation of a sitting president would at least disturb legislators called upon to "judge" him. The effect on the Colombian legislature is completely different: it spurs, in the name of patriotism, an increase

---

*With no more aid from the United States, a major crisis in the diplomatic relations between the two countries occurs. Colombia's government is cast out because of the alleged ties between our president and the Cali cartel.

in solidarity and team spirit among most of my fellow legislators, who are themselves closely connected with the mafia.

But on March 5, one event shows me just how shaken the mafia is by what is happening. José Santacruz—the Cali kingpin, the man who came with the Rodriguez brothers the day I talked with them—is killed. He's killed while trying to escape, we're told. He was imprisoned along with the three Rodriguez brothers. Why did Santacruz suddenly decide to break the silent pact by escaping? To me, his act proves that the bosses of the Cali cartel are becoming aware that they've allowed themselves to be trapped by the president. If they help him out, as they've been doing up to now, he benefits from it. If they say anything bad about him, that does him a world of good, too. In short, they no longer have any chips to cash in, and they no longer see how they're going to negotiate their silence with this pressure the United States is putting on the government. As far as I can see, Santacruz was the first one to lose patience, and he paid for it with his life.

During this month of March I'm slowly recovering from my hunger strike. But the official date for the opening of Samper's "trial" is approaching, and I'm beginning to rally my troops. My friend Clara, who campaigned with me for the House, offers her help. She's a lawyer, and I don't yet realize how valuable her aid is going to be. We meet at my home, like in the good old days, and there we are again redesigning the world, and Colombia in particular, over a cup of coffee and a conscientious overview of the newspapers.

"They say that legislators can get Valdivieso's complete investigative file. Why don't we take them at their word?"

"There are thousands of documents in that file, Clara; it's completely unrealistic, but you're right, let's beat them at their own game. Just see how guileless we are, they say, their hands on their hearts. Well, okay, my friends, we're going be guileless too, but we're going to go all the way! All the way!"

Valdivieso's file is placed at our disposal. The president's office, trying to convince Colombians of its sincerity, has gone so far as to have the file published in the official journal.* But it is in jumbled fragments, in an inextricable disorder, so that no one ever notices what is missing, so that it's impossible to find any coherence in it. Most people would give up trying to read it for fear of going mad. Not Clara. Clara's used to this kind of puzzle. The pieces pile up, they occupy the whole dining room table, overflow onto the carpet, and soon fill the vestibule and the bedrooms, but Clara doesn't give up, and neither do I. Scissors and stapler in hand, we reconstitute the depositions, start a file for each witness, reestablish the chronology, and, with an unimaginable persistence, succeed in restoring clarity and coherence to what was at first only a mass of papers that had cleverly been made unusable.

We spend days and nights sorting it out, but when we've plumbed the depths of this unprecedented, giant labyrinth, I can see what a miracle we've just accomplished: the file is explosive, not only for Samper but for the whole Colombian political class. Everything is there, far beyond what we might have hoped. Alongside testimony extremely damaging to the head of state, we find the secret accounting from his days as a candidate, proof of the sums he paid to elected officials to ensure their support and, through them, to give bribes to millions of impoverished citizens in order to obtain their vote. The full extent of the system of pyramidal corruption peculiar to Colombia is unveiled for the first time, with hard facts and figures to support the revelations. The elected officials have signed receipts for candidate Samper, and these receipts are laid out before our eyes. And what do we find? That most of these representatives who are now preparing to "judge" Samper have signed such receipts. These men who loudly

---

*Gaceta Oficial is printed by the state to disclose all congressional activities (debates, new laws, amendments, etc.).

proclaim their concern for fairness and ethics have illegally taken money from the man they claim to be challenging. How can they judge him, when Samper has a hold on them, when in fact they're his accomplices? There is enough there to dynamite the trial and show Colombians the rotten underbelly of our political system.

My task is to find the words to turn this into something that will "speak" to the millions of people who will be following the trial on their television screens. This is a complicated, almost impossible task, as it's difficult to make a resounding appeal using accounting data. I've understood that once he's found not guilty, Samper's strategy will be to say, "Look, they've had plenty of time to go through the documents in the prosecutor's file, and they haven't found anything against me. Isn't that the best proof of my innocence?" I want to throw a monkey wrench into the works, to divulge his game.

The trial begins on May 22, and as one might expect after the United States' "decertification" of Colombia, all we can hear are ringing appeals to close ranks around our president, whom the gringos would like to bring down in order to replace him by one of their own men and humiliate Colombia. Has Samper gotten wind of the considerable labor Clara and I have accomplished? Since my hunger strike, I have the almost palpable feeling of being followed, spied upon. Later on, I find proof that, in fact, policemen took turns reporting everything I did and that my telephone was tapped.

I'm scheduled to speak on June 11. Almost two weeks before, my concierge hands me the usual pile of mail. I'm coming home from the legislative chamber. It's almost eight in the evening, and I'm preparing to work far into the night on still another version of my public presentation. In the elevator—I live on the eighth floor—I take a quick look at the envelopes: bills, ads, and, hmm . . . a handwritten letter. I open handwritten letters first because they're generally friendly and personal. I take out my keys, step into my apartment, and turn on the lights. The apartment is quiet. This evening, Melanie and

Lorenzo are sleeping at their father's place. Something falls out of the envelope as I pass through the vestibule, and I mechanically pick it up, not really looking at it, and begin to read the letter. It's a page of crude insults, insults in which the last paragraph knocks the wind out of me: it says that now they're going to make sure that my children pay for what I'm doing.

This is something that has never happened before. In my earlier battles against corruption, my children were never threatened. Then I think to look at what I picked up a moment earlier. It's a photograph, showing the body of a child cut to pieces.

At first, anger overcomes fear, and I shred this horror, angrily, stupidly, before crushing it under my heel at the bottom of the garbage can. In order to regain control over myself, in order to be able to breathe and live a few more seconds as if it hadn't existed, as if I hadn't seen anything. They won't intimidate me. This case is closed. And then, on second thought, I tell myself, others who have been too close to this scandal have died under mysterious circumstances. These past months: Serpa's driver, Álvaro Gómez, Elisabeth Montoya, José Santacruz. Those dead faces, photographed by the police and widely distributed by the press, pass before my inner eye. Oh, my God! I go back to the garbage can. Call Fabrice, right away!

"It's me! Are the children with you?"

"They're eating dinner. Do you want to talk to them?"

"No, don't tell them anything. Look, we have to see each other, Fabrice. It's very urgent, I can't talk to you on the telephone."

"Are you in trouble?"

"Yes. Call me as soon as they've gone to bed and I'll come by."

We live not far from each other, near the French embassy where Fabrice works and the French school the children attend. Melanie is ten and Loli will soon be eight.

"We'll send them away tomorrow, Ingrid. I'll call my mother, and she'll pick them up at the airport."

Yes, that's right, tomorrow. Not another day in this nightmare. That night, we pack their suitcases. We alert the French embassy. An escort is arranged for first thing in the morning. The children will take a plane to Paris in the afternoon. They'll go to the embassy first thing in the morning, where they will stay, under protection.

When Melanie and Lorenzo wake up, they're a little confused. How can we explain that they have to go away, leaving their books and notebooks behind, without even saying goodbye to their teachers or their friends—in short, that they have to run away like fugitives?

They're satisfied with vague pretexts. For now, I think, you will be safer in France. Because of what's happening in the Colombian legislature and, of course, the grotesque letter, I'm too agitated to talk. "Let's go, let's go, my dears," I say, "we have to go quickly, the people at the embassy are waiting for us, we'll discuss all of this in two weeks. And give Grandma a kiss, she's so happy you're coming."

It's a relief to know that they'll soon be at their grandmother's home! I'm already imagining her hugging them as they step off the plane, the streets of Paris luminous under the June sun, their children's rooms in the old, distant, provincial city where Fabrice's mother lives. I'm even able to muster an inner smile. Yet their departure sounds a first alarm bell in my life as a woman, as a mother: for the first time, my political activity is having concrete and serious repercussions on my family. Up to now, except for Lorenzo's refusal to eat during my hunger strike, I've largely succeeded in keeping my family separate from the unheard-of violence of public debate in Colombia. From now on, I can no longer ignore the fact that whatever I undertake in my public life is fundamentally affecting the existence of those who are closest and most precious to me.

But for the time being, the departure of Melanie and Lorenzo increases my will to fight. This time, other than killing me—and they won't kill me—they can't do anything to stop me from speaking. We're now a week away from my speech, and the pressure is great.

Throughout the country, many people are waiting to hear me. Samper's so-called trial has become Colombians' favorite soap opera, and from five in the afternoon until late at night they remain glued to their television sets, following this soap opera, harboring a crazy hope that someone will finally say out loud what's being whispered on every street corner. Many wish it will be me. My rendezvous with the Colombian people haunts me. I have no expectations regarding the outcome of the "trial," but I'm convinced that Colombians can overcome yet another deception if I succeed in giving them a reason to believe in ourselves as a nation.

"Juan Carlos, you have to find me a sign, some kind of logo that will send an instant signal to the people. I have to establish a bond with them right from the beginning."

Juan Carlos, an architect who has specialized in advertising, is my closest advisor during these feverish days. Sensitive and discreet, he has a marvelous ability to listen.

"I understand. Let me think about it, and I'll make a couple of suggestions tomorrow."

The next day, he brings me a drawing of an elephant, and I burst out laughing. Of course! For days people have been talking about the scathing remark made by the archbishop of Bogotá when asked by a journalist whether Samper could really have remained ignorant of the considerable illegal sums the Cali cartel has sunk into his campaign: "Listen, when an elephant comes into your house, it's difficult not to notice it, isn't it?"

"That's brilliant, Juan Carlos! Brilliant! I've got to wear this elephant on June 11."

"I'll take care of it, the elephant is all people will see, don't worry."

All that remains for me now is to get the legislature to assign me the prime-time slot: between seven-thirty and eight in the evening. A priori, it's hard to see why the leadership of the House, composed of strong Samper supporters, would do me this favor. On the morning of June 11, before I've even left home, my musketeer friend, Maria Paulina Espinosa, tells me that I'm at the top of the list of those who are to speak, scheduled for noon. Why not just give me a first-class burial? They think they can get away with this? We'll see about that.

I decide not to show up in the legislature before four in the afternoon, and to disappear in the meantime. Since the entire press corps is announcing my speech as the event of the day, they'll be forced to let me speak at whatever time I choose.

Some of my colleagues are sent out to look for me. Everyone panics. The more time goes by, the clearer it is that I'm going to be on television at the worst possible time for them. In desperation, the president of the House adjourns the session. That suits me fine; when it resumes, I'll be the first speaker. My arrival is greeted with hysterical reprimands in the corridors.

"Who do you think you are? Do you think the whole country has to adapt to your schedule? People don't give a damn about you."

"Well, all the better for you if they don't give a damn about me, but I won't speak before seven o'clock, that's how it's going to be."

The leadership is furious. But the press is everywhere, and excitement is at its peak. They can't do anything to countervail.

"All right, you'll speak at seven, but not for more than an hour."

"Sorry, I'm speaking at seven and I have three hours' worth of material. Up to this point, everyone has had all the time he wanted, and there's no rule that allows you to limit the length of my speech."

Representatives give me vicious looks, slam doors. It's strange how their angry gestures, their hateful words, amuse me instead of hurting me. After these weeks of insomnia and fever, I suddenly have the feeling that I'm floating in a cloud of bliss. I think about my fellow citizens, with whom I'm conducting a sincere dialogue about the problems we share, and with whom I'm constructing a future in spite of everything, over the heads of a political class steadily being chipped away by disgrace.

It's precisely seven o'clock when I mount the podium. To stand out clearly against this assembly of flushed and furious men, I'm wearing a light blue miniskirt and a jacket of the same color over a simple T-shirt. All the cameras turn toward me. When I take off my jacket, people see Juan Carlos's elephant printed on my T-shirt with this statement in capital letters: ONLY THE TRUTH! The tone is set, and on every face is consternation—a hostile but powerless consternation.

At this stage, I take the opportunity to point out that the T-shirt Colombians have just discovered on their television screens is about to become a symbol. In the streets over the next few months, wearing it will be a way for people to express their severe reproof of President Samper and of all those who claim that this comic opera was an honest trial.

I then begin my address by talking about what we are going through as a nation. In a very simple way, as if I was explaining this to Melanie or Lorenzo, I talk about myself, my experiences, and my feelings over the past months as truly as I can.

"I used to be Samper's friend. You remember how I supported him in his campaign. When I heard the rumor that he'd been financed by the mafia, I laughed. Come now, are the enemies of our young president losing their cool? Who could believe such allegations? Samper talked about a mafia conspiracy, and I applauded him, like a good little soldier. That devilish mafia that doesn't respect anything, not even the Nariño palace! What, now the Americans are falling into this trap? They're threatening to decertify Colombia? That's just because our independence scares them, Samper calmly announces, and once again I applaud loudly. But now people start talking, and disturbing documents are turning up here and there. I still want to believe our president, but our president is suddenly no longer talking about a conspiracy. Now he says that, on closer inspection, it appears that the mafia might in fact have financed his campaign, but he didn't know about it. His closest advisors are supposed to have known, but not he. He's pure as the driven snow, he spent these millions of pesos without having any idea where they came from. I tell you, even a dope like me found that a little difficult to swallow. Especially since more documents were appearing. Look, I've got them right here, and I'll show them to the camera so you can also see what they look like."

For hours, I present to the Colombian people the irrefutable proof of the president's guilt. We work our way through the labyrinth I reconstituted with Clara, which is composed of testimony, receipts, photographs, letters, speeches. I'm telling Colombians a very dark story, the story of a man who has decided to do anything in order to become president of the republic, who'd been connected with the Rodriguez brothers for years, who was surely convinced that he was acting in accord with the rules of an excessively permissive society, and who was himself completely unscrupulous.

"As far as I'm concerned," I go on, "the discovery of these documents was a terrible blow. Fortunately, we live in a democracy, and while the president is clumsily trying to justify himself, justice is following its own humble path. A president who lets the judicial system do its work might deserve the benefit of the doubt—unless he tries to manipulate the process, going so far as to buy the investigative commission. I really wanted to continue believing in this man! However, when the first person was killed, my blood ran cold. Who killed the minister of the interior's driver just as he was getting ready to testify? Who had an interest in silencing that man forever? The police, oddly enough, never arrested a suspect. Did they even search for suspects? It doesn't look like it. And then, one after another, the witnesses are gunned down by mysterious killers. A little while ago I showed you the proof that our president is a liar. But that's not the most serious thing. What's more serious, you see, is that I'm now convinced that our president is a delinquent."

A deathly silence falls on the House. Such words have never been uttered here. I can hardly breathe, but I go on, choking with emotion:

"You can imagine how hard it is for me to make such accusations. Because it's a heavy burden to be alone in bearing the truth in this masquerade that people are telling you is a trial. Within a few hours, these representatives you see looking a little stunned are going to exonerate the president. Why? Because they know that by saving him they'll be saving themselves. You see this man sitting here in front of me, for example? Well, I have a receipt signed in his own hand. He has received millions of pesos. I tell you, looking you straight in the eye, if he ever made up his mind to vote against the chief of state, I'm not sure he'd still be alive tomorrow morning. We Colombians are powerless spectators watching a play whose outcome has been determined in advance. This evening, our country is in the depths of an abyss, in its death throes, and yet I know that the day will come when our aspiration to happiness will win out over the dizzying attraction death has so long exercised over us. I have confidence in us, the people of Colombia."

As I return to my seat, something curious, and unusual in a Colombian context, occurs: the silence continues, an impressive, stupefying silence, as if these men, who are only too prone to violence, are temporarily broken. Some of the representatives on the conservative side rise to shake my hand, but without saying a word, apparently paralyzed, stunned.

That night, around two in the morning, Ernesto Samper is officially exonerated by a vote of 110 to 43.

# CHAPTER EIGHTEEN

**FROM THAT DAY ON,** as if to show me that the legislature's vote did not reflect that of the people, men and women come up to me in the streets every time I go out and hug me or offer me a word of encouragement: "We're there, Ingrid, you have to hold on, keep going." In the next days, we manage to hand out more than five thousand T-shirts with Juan Carlos's elephant on them, clearly defying the verdict.

When the legislative session ends eight days later, the first thing I want to do is join Melanie and Loli. Now that the excitement of the debate is over, guilt consumes me: it's my fault that they've been exiled from their familiar world. At the end of June, dreaming of family breakfasts, walks in warm summer breezes, long afternoons, Juan Carlos and I fly to France. I'm exhausted, my nerves stretched to the breaking point. Five months have passed since my hunger strike, and I haven't had a single day of rest.

We've been together only a week, and we're just beginning to recover the sweetness of normal, everyday life, when a phone call from my lawyer plunges me into the depths of anguish. Hugo Escobar Sierra, the gray wolf of the Colombian bar who got me through the Colt affair, is obviously very worried.

"You have to come back right away, Ingrid. I'm really very sorry."

"But why, what's happened?"

"They've indicted you, my poor child. For influence peddling."

"What!"

"This is coming directly from the presidency, from Samper. It could be serious; you could lose your seat in the legislature."

"But that's impossible! What influence peddling? I've never asked for anything, never solicited anyone."

"I can't tell you anything on the phone. Take the first plane. I'll be waiting for you."

This is a terrible heartbreak for the children, and for me. We'd planned a trip, made hotel reservations, and now everything is going to fall through. Fortunately, Fabrice's mother is there, available and generous. She takes them in. And Juan Carlos, efficient and discreet, is booking us two seats for Bogotá, at the beginning of the summer season, the worst time of all. At times like this, I think that without him I'd lose faith.

The flight back seems endless. In the plane, I can't eat, I can't sleep. How have I been guilty of any kind of influence peddling? If the president's office thinks they can make this charge plausible, they must have something that looks like evidence. Desperately, I rack my brain: What could it be?

Hugo Escobar Sierra has not exaggerated the gravity of the charges against me. The proof is already there at the airport, where journalists are waiting for me. Unable to answer their questions because I still don't know the charges against me, I leave through a side door, escorted by men from the airport's security service.

"You met with Ernesto Samper at the beginning of his term," Hugo Escobar Sierra begins in a very calm voice, his hands lying flat on his desk. "Do you recall this meeting?"

Yes, at that time everyone was talking about the cassette recorded by the Americans, on which the Rodriguez brothers were heard praising Samper. And the president, as I've already mentioned, said to me: "Don't talk so loudly, Ingrid. The gringos have bugged my office."

"I remember it perfectly. It was a formal visit, at the beginning of his term."

"What did you talk about?"

"Nothing. He was receiving all the legislators who were part of his majority. It was a matter of touching base, that's all."

"No. The president says that on that occasion, you asked him to do a favor for your father."

"What! That's false, I didn't ask for anything at all!"

"Ingrid, did you or did you not mention your father's situation?"

Then I recall that in fact, I did refer to my father's difficulty in getting along on his pension, which hadn't been raised in twenty years.

"I remember now, yes. He asked me how Papa was doing and I had to say that he was having money problems."

"There you go! Well, that's all it takes to send you before the Council of State."

"But that's crazy! I'd never have mentioned Papa if Samper himself hadn't asked about him."

"Wait a minute, Ingrid, there's something involved here that I don't understand. Why did Samper ask you about your father?"

"Because Samper is an old friend of my parents!"

"He is? Why didn't you tell me that before!"

"I thought—"

"This changes everything, Ingrid. If you can prove that your father's situation was mentioned for obvious reasons of mutual friendship and not as a request for a favor, you have a chance of changing the judges' minds. Otherwise it'll be the president's word against yours, and I have to tell you it's going to be difficult."

"How can one prove friendship?"

Do everything you can to find me the slightest evidence of a connection between your family and Samper. Do the impossible, Ingrid. And do it quickly—we have very little time."

I go home distressed, panicked, and call the only person who may be able to help: Mama.

She listens to me, and then says:

"If that's the way things are, I'm going to tell you a secret, Ingrid."

Then she tells me under what circumstances she and my father became connected with Samper. When Papa was Colombia's ambassador to UNESCO and we lived on Avenue Foch, Andrés Samper, Ernesto's father, had a subordinate post at the embassy. He was a fragile man who drank a lot and had money problems. One morning when he was depressed, he slit his wrists in his bathtub. The person who found him, perhaps the concierge, tried to find someone to call and came upon my parents' phone number in his address book. My mother rushed over. It was she who gave first aid to Andrés Samper and took him to the hospital. In the days that followed, she was very attentive to him, as Mama usually is with people who are suffering: spending hours by his bedside, protecting him, comforting him. So when the future president's father left the hospital, he agreed to convalesce at my parents' apartment on the Avenue Foch. Mama had a sincere affection for him, full of compassion. He spent two months at our place, where I undoubtedly met him without retaining the slightest memory of it. Then he was called back to Bogotá. After that, my mother maintained a friendly long-distance relationship with him by correspondence. When Mama returned to Colombia and got back into politics, she was in the Liberal Party, just like Ernesto Samper, Andrés's son. So they weren't just friends; they worked together until the day my mother decided to join Galán.

I'd always wondered about the strange friendship between Ernesto Samper and my mother. Now I had the explanation.

"Fine," I tell her, "but I've got to have proof of this story."

"Proof? Wait a minute, I might have what you need. I recall that at Andrés' death I sent Ernesto a note expressing my condolences, and he replied. A letter, I think, in which he mentioned that period. I believe I kept it. Listen, give me an hour or two."

Mama hangs up. Half an hour later, the phone rings.

"I've got it, dear. It's dated April 18, 1988. I'll read it to you: 'Dear Yolanda, thanks very much for your generous note at the time of my father's death. He never forgot the love and help you gave him during the difficult times he experienced in Paris. Please accept on behalf of my family and myself our most affectionate and warmest embrace. Ernesto Samper.'"

The day of the trial arrives. Once again I face the pack of journalists, and then the grim faces of my judges. People say the Council of State is pro-Samper, and suddenly I feel terribly scared. How can I fight all alone against a whole state apparatus? My fear must be visible in my face, because my elderly attorney tenderly takes my hand.

"Courage, Ingrid. It's going to be all right, I'm here."

Ernesto Samper's complaint is apparently flawless. He claims he doesn't know me, that as far as he's concerned, I'm a legislator like any other. He was therefore astonished by my request to benefit my father. Then he goes even farther. He hints that if I was the only one to oppose him during his trial, I did so out of spite, because he refused to help me get an increase in Papa's pension. He claims that up until the last moment—that is, until the opening of the trial—I'd harassed him in order to obtain his help, but all in vain; hence my anger and my determination to hurt him.

The journalists are excited, of course. So that's Ingrid Betancourt's secret motive: revenge! That's what all these fine speeches about ethics concealed! Really, she's no better than the most corrupt of the politicians.

But by trying to prove too much, Samper has made a mistake, and that's where I launch my counterattack. I'm supposed to have harassed him? If this is true, how can he explain that I am the only member of the legislature who has systematically refused all invitations to meet with him—since 1995, that is, almost a year before the trial began? I kept copies of some of the stinging replies I sent him, in particular this one, dated April 1996, which I read. "Mr. President, I thank you for your invitation to participate in a little working breakfast on the 25th of this month, an invitation I must refuse because we will not have concluded our investigation into the charges on which you are to be tried. Cordially, Ingrid Betancourt." A kind of dismissive reply for someone who is thought to be harrassing the president.

Then I come to the heart of the matter, the friendship that binds my parents to the Samper family and explains why the president politely inquired about my father. Before a hypnotized audience, I recount the tragic story of Andrés Samper. In conclusion, I read the future president's note to my mother.

"After that," I say, "how can the head of state claim that he doesn't know me?"

And for the benefit of my judges, who now seem to be avoiding my eyes, I add, in a lower voice that reflects my sudden discouragement:

"Cooking up this kind of accusation against me is shameful. It's so difficult in this country to conduct a serious political opposition, and how can one succeed if one has the judicial system against oneself?"

Then the representative for the prosecution, a rigid, sour-looking woman, stands up, and I expect the worst. My stomach is tied in knots, my head is spinning. I lose the thread of her argument; I'm incapable of concentrating on anything except the bottomless despair that's overcome me. But suddenly it seems that I'm dreaming. What?

What's that she's saying? That the Council of State should never have gotten involved in this kind of masquerade? That the file contains nothing but malicious gossip whose inanity Representative Betancourt has just amply demonstrated. I lift my eyes. My attorney smiles; he's also astounded. The journalists are looking at us in a completely different way now, with good will. Am I really winning, really? "That is why I ask you," the little lady concludes angrily, "to immediately close this trial and never mention it again!"

It's over. Everyone gets up, and, while the press literally pounce on us, three of my judges, three women, distancing from their peers who are leaving the hearing room and walk towards me to shake my hand. "I wanted to tell you how close I feel to you," one of them murmurs.

On July 20, 1996, four days after this trial from which I emerged exhausted but uninjured, the legislature reopens its doors. Traditionally, at this time, the House and the Senate meet jointly to hear the head of state's address. This is a day of glory for Samper: five weeks after the closure of "his" trial, he returns before the elected representatives of his country, holding his head high. For me, it's a day of mourning, an unforgettable day. Theoretically, I'm supposed to be seated among the Liberal representatives, but since all or nearly all of them supported Samper, I refuse to sit next to them. Anyway, most of them look away when I pass by. This time, I'm going to sit at the back of the chamber.

Suddenly there's a sound of stomping boots, orders being issued. The president enters. We rise. Ernesto Samper solemnly mounts the podium, salutes, and then freezes. The first notes of the national anthem sound. Is he looking for me? Maybe. In any case, I caught his eye and now we're looking straight at each other over the heads of the legislators standing at attention. The image of this man joking pleasantly during our trip to Maicao, ten years ago almost to the day, comes back to me. There is nothing comical or charming in the way he looks at me now; I see nothing in his eyes but a deep, steely, implacable hatred. But at this moment, I want him to know that I'm judging him in the name of the victims of his insane ambition. In my heart, I am struck by the conviction that this is the source of the death threats to my children. I communicate to him in this way, silently, and an exchange of unimaginable violence passes between us while the national anthem seems to hold time in suspension. But I also want him to know, and this is the essential point, that I have no hatred for him, that I don't give a damn about the emotions, the pathos, the personal feelings that inevitably affect both of us. I want him to understand, if he's capable of understanding, that I find him guilty with regard to our people, with regard to our history, and that this is far more important than the sinister test of strength in which we're engaged.

On this July 20, immediately after his address, Ernesto Samper is giving a cocktail party in the presidential palace in honor of the legislators. The Capitol building is separated from the Nariño palace only by a garden. That evening, as I see all my colleagues joyously heading off through the garden, I feel profoundly depressed. Before the eyes of the Colombian people, the legislators have done their best to bury the truth and save their skins. Now they will unscrupulously receive their rewards.

I set out in the opposite direction, toward the Bolivar Plaza. It's a very cold night, one of those nights when the icy wind off the Andrean mountains empties the streets of Bogotá. The plaza, so lively during the day, is deserted. Only my personal car and that of the bodyguards whom the ministry of the interior has recently assigned to me are parked on the right side of the plaza. I go down the steps of the Capitol building and then hurry toward the car. I have complete confidence in my driver, Alex, but I have less confidence in my bodyguards, who belong to the state police. I've just managed to arrange for them to have their own car, so that they no longer travel in my own, where I suspected them of listening in on my conversations. As I turn up the collar of my jacket, Alex spots me; he starts the car and turns on the lights.

We go up the street on the side of the cathedral, then turn right and enter San Felipe de Neri Street. The escort follows, its headlights illuminating the inside of my car. The narrow streets of the historic Candelaria quarter, the most picturesque in Bogotá, are completely deserted. But oddly, when we try to turn left in front of the university, a car blocks our path. I hear myself say:

"He's a jerk, that guy! He has all the room he needs, why is he stopped there?"

My driver also seems puzzled. We can't get through.

I turn around impatiently. I'm tired, depressed, and I want to go home. The best solution is to back up and take the preceding street, but I see that another car has blocked our retreat.

"Damn! We can't back up, either."

Suddenly Alex understands.

Fortunately, I have a little car with four-wheel drive. Alex turns the wheel sharply, guns the motor, and drives right over the sidewalk and into the narrow alley that remains open between the corner of the university building and the car that is barring our way. He gets through and roars at breakneck speed up San Francisco Street. I finally understand as well. We're already far away, almost at the lights of Septima Avenue, Bogotá's big artery, when we hear shots ring out. Then I realize with relief that the escort has also gotten away.

We've had a close call, and it's clear that these people, who were paid to shoot at me, will try again. My first instinct, however, is still to believe that they didn't intend to kill me but only to intimidate me.

A little farther on, I have both cars stop.

"Not a word about what just happened. All right? Not a word, I don't want anyone to know about this."

My bodyguards will talk, of course, but it won't go beyond the police department.

I don't want to see the reality of this, and in my heart I know why: if my family finds out, if Fabrice finds out, Melanie and Loli will never come back to Bogotá. They're scheduled to return at the end of August, and I live only for that day.

# CHAPTER NINETEEN

**AT THE BEGINNING** of September 1996, Melanie and Lorenzo go back to the French school in Bogotá, exactly as if nothing has happened. For them, however, this school year will be very different from the others. Their father is no longer living in Colombia; he's been posted to Auckland and has made the move over the summer. From then on, the children have only one home, mine, and if something happens I know I can't call on Fabrice for immediate action. But we've gained a refuge, a sanctuary; if life in Bogotá becomes impossible, they'll have Auckland. Neither Fabrice nor I have clearly formulated it, but obviously we are thinking about it. Since I received the threatening letter with the horrible photograph of the dismembered child's body, it pains Fabrice to think of Melanie and Lorenzo in Bogotá. At the same time, he doesn't think he has the right to take them away from me. I've appreciated his concern and his thoughtfulness. For my part, I'm aware of the risks to which I'm exposing them, but separating from them is just beyond my strength. I feel that I've given a great deal of myself to this country, to politics, but I'm not yet ready to sacrifice my family life.

So, for me, the autumn of 1996 begins with an unspeakable fear. In order not to lose Melanie and Lorenzo, I've concealed the attack on July 20 from which I barely escaped. Now I'm paying the price of my irresponsibility: I'm constantly afraid for them. It is a primal fear, an octopus whose tentacles are tightening around my stomach, tearing at my heart. I carry it within me, I can't forget it for a single instant—not even when the children are both there, near me, in their pajamas, in their room, and Juan Carlos is checking the locks on the entry door. Even at these moments, which should be so sweet, I'm afraid. I feel guilty.

I've withdrawn from the world in order to wage my final battle against Ernesto Samper: I'm writing a book in which Colombians will find all the proof of his guilt. Solitude, I learn, can amplify fear, and I feel it at every instant as I listen for any unusual noise. And yet I keep writing, obsessed by the need to get on with it, to salvage whatever part of the truth can still be saved. I'm writing so that the "trial" of Samper, aborted by the representatives, can still take its place in our history, so that we will never forget the indignity to which we have been subjected. They deserve to have, in black and white, all the damning documents that I held up before their eyes three months earlier, during the trial in the House, before the television cameras. Our history will still hold true.

I recall a lecture by Hélène Carrère D'Encausse at Sciences-Po in which she gave a moving description of how totalitarian regimes manage to rewrite history: collective amnesia, tailored to order. Modifying events, changing outcomes, altering official photographs, making people who are out of grace simply disappear as if they never existed. That horrified me more than anything. It seems to me that the cowardly relief we'd all be tempted to feel would seal our fates. At such dark moments, so heavy with threats, I tell myself that if there's one battle I still have to fight before they kill me, should they kill me, it's this one: to prevent this final manipulation, this ultimate humiliation.

I hardly ever go out any more. I barricade myself at home, and I write, night and day, feverishly, in the mad hope that this book will exist as an assurance of salvation for all of us, Colombians, but also for my own family, for Melanie and Lorenzo. They're so vulnerable—who will protect them? How? How could this book protect them? I can't say. But an intuition, a conviction, that this duty has to be done and that it will give me the strength to go on.

I can't sleep. The moment I lie down, my mind constantly replays the scenario for our escape in case we are attacked. My apartment building is at the end of a cul-de-sac, backed up against the mountain: an ideal place for an ambush. If the killers come up the stairs, how can we get away, how can we save the children? At first, I think of using a knotted rope to reach our downstairs neighbors' balcony. But going down the rope is itself a scary proposition; a rope ladder seems so much safer. Well, all right then, a rope ladder—I will go order one tomorrow morning . . . When the ladder is there, ready for use, I start to worry about tennis shoes—we all have to have our own tennis shoes, and they have to be kept at all times in front of the sliding glass door to the balcony. I don't want the children to be going down that ladder in their bare feet; they could slip. Oh, no! My God . . . what if they come down from the roof? We've got to have a gun, we just have to. We'll find a pistol, a submachine gun. Yes, that's it, I'll learn how to use a gun, Juan Carlos will teach me. We'll hang the gun up to the left of the entry door. Lorenzo, you're forbidden to touch this gun, do you hear? It's very dangerous, very dangerous. What's happening to me? I'm delirious. Am I going mad?

The book comes out on December 12, under the title *Sí Sabía*, "Yes, He Did Know." Ernesto Samper did know that the millions of pesos poured into his election campaign came from the mafia, I assert. On the day the book is published, a signing is organized in a downtown bookstore. Will people dare to come? Will they dare to brave the press, which has announced the event, and take the risk of being identified in photographs and on television? We're afraid of an attack, I'll be signing books at the rear of the bookstore, with my back to the wall. The two entrances are guarded by armed men, and other guards are posted discreetly throughout the interior of the store. But when the people come in, they crowd through the doors with an audacity and a pride in being there that soon make us forget our security arrangements. There's a throng; the line stretches far down the sidewalk. Many have brought cameras and ask permission to have their picture taken with me. They put their arms around me, kiss me. I feel as if I'm coming back into the light, tasting, for the first time in months, the enormous pleasure of living. These people are liberating me, opening the doors of my prison, telling me that they've heard me, that they know, that they support me. I'm so moved, so overcome by their confidence that I'm surprised to find myself once again believing that we can win, that Colombia will not always be run by thieves, crooks, and criminals.

In the days following the publication of my book, I resume my activities in the legislature. I've had to neglect my legislative work, when I was writing the book, but now I work overtime. The session is almost over; with the long Christmas vacation approaching, I have to act quickly. I rush back and forth between the legislature and my office, where I have one appointment after another.

On one of these crazy days, as I'm trying to wind up my appointments before running off to the plenary session, my secretary pokes her head in the door of my office.

"Someone's asking to see you right away, Ingrid. A man."

"Does he have an appointment?"

"No. But he's very insistent."

Later on, I wonder why the people who have sentenced me to death sent this messenger to warn me. Then, recalling my meeting with the Rodriguez brothers, I figure it out. I sense that deep down, they may not really want me dead; what they want, on the contrary, is for me to continue to exist in spite of everything. For people like them, who buy everything and everybody, truthfulness is the only value they still respect and admire. It is the proof of courage and freedom. It is a sign that not everybody has a price. This implies that it is possible to believe in a world of justice, not of arbitrariness. This is perhaps naïve, but I feel that if Colombia could have given those guys true opportunities perhaps they wouldn't have chosen to be criminals. Even for them, Colombia has become just hell. I suppose, from what I heard from them, they expect another kind of life for their children. This means, in a way, a hope that one day Colombia will have rules that will apply to everybody, not only to those who can pay up. You won't have to be born rich or become a criminal to have justice or to have impunity, both of which you can buy today in Colombia.

"Your family is in danger," the stranger says. This time I can no longer play games. That night, we prepare to depart.

For the second time in six months, Melanie and Lorenzo will leave Colombia quickly. But this time I'm sure they won't come back. For a long time. The promise of this departure relieves me after the months of fear and guilt. I don't want to think about what my life will be like without them—not yet. Juan Carlos is there to comfort me; Fabrice is waiting for us in Auckland. For the time being, in the eye of the storm, we're hanging on. Sebastien is there as well. He's now a man and, above all, an attentive big brother, very fond of Melanie and Lorenzo. He has also experienced comings and goings, trips to the ends of the earth, difficult changes. He understands. He takes care of them, like a guardian angel.

When I think back on the two months spent in Auckland, the last in my family life, I can't imagine myself in a role other than that of a woman condemned to a long prison term who's feverishly taking advantage of her last days of freedom—to arrange for the daily life for her loved ones, attend to the tiniest details, even completely insignificant ones, in the mad hope that her absence will not cause them suffering—or, rather, perhaps they won't forget her. Together we decorate their rooms, try out their route to school, buy books and school supplies, go shopping for school clothes for the whole coming year. "Take the next larger size, Melanie, you're going to grow, you know." Build up a fund of memories. And in my heart, I say to myself, "Look at her carefully now, Ingrid, otherwise in a few months you won't be able to remember how she looks. Take a good look at her, register everything, the way her pants are cut, the color of her blouse."

In the middle of February 1997, Juan Carlos and I return to Bogotá. We both sense how difficult things are going to get, and Juan Carlos chooses this precise moment to do something entirely characteristic of him: he asks me to marry him. How better to show me that he is with me—in love with me, but also in solidarity with me, prepared to follow me to the end of the road? We've both been married once before, and in ordinary times we wouldn't think this merely formal bond important. But we're not living in ordinary times, and Juan Carlos's words bowl me over.

We're married on the way back, in the middle of the Pacific Ocean, according to the Polynesian ritual. A biblical marriage, outside time, where the future husband slowly emerges from the waves in a dugout canoe, as if we've been allowed, for just one day, to go back to the origin of life, before human beings lost their innocence. Just for a day or, rather, for three days, until a plane picks us up on this archipelago, sung by poets, and carries us back into the furious convulsions of real life, our life, in Colombia.

Ostracized by other nations because of its president (Samper will be denied a visa for the United States when he tries to attend a United

Nations session), the Colombia we return to in 1997 is already looking forward to new elections and thus new hope: the presidential
elections are only a little more than a year away. In Colombia, a president serves a four-year term and cannot be reelected; thus we're sure
to be rid of Ernesto Samper soon. But are we also sure we will recover
our dignity? No, because Samper is seeking to protect his way out. He
knows that once he's an ordinary citizen again, he'll no longer be
sheltered from prosecution. The trial 8000 is still being pursued
against his closest collaborators, notably Medina, his treasurer, and
Botero, his defense minister. So while he's still in power he has to try
to throw a monkey wrench into the judicial inquest that will end it
for good. Basically, if he wants to live in peace with the fortune he's
accumulated while in power, he has to impose his own successor on
Colombians. This successor's name is already on everyone's lips:
Horacio Serpa, the minister of the interior, the most faithful of the
faithful and as corrupt as his mentor. There's no doubt that if Serpa
succeeds Samper, the latter will be able to live at ease.

Powerless, I watch as the trial 8000 is sabotaged. And why am I so
powerless? Because the strategy Samper is using is not vulnerable to
any criticism. The man who keeps the trial 8000 going is the prosecutor, Valdivieso. And what is Samper doing? Backstage, he demands
that the Liberal Party designate Valdivieso as its candidate for the presidency! Few men could find the courage to resist that kind of promo-
tion, especially if they're told that, according to polls, they'll win by a
landslide. Valdivieso succumbs to the siren song and resigns his office
in order to enter the race for the presidency. That's his right, but he
fails to see the trap Samper is laying for him. A poor speaker and a
mediocre politician, Valdivieso, in a matter of months, will lose all the
credit he's accumulated in the minds of Colombians and will disappear from the scene. Then the real candidate of the Liberal Party,
Horacio Serpa, will emerge into the spotlight. Meanwhile, Samper
will have succeeded in nominating as Valdivieso's successor one of his

most devoted attorneys in the trial 8000! This is a checker master's move. It's easy to see what will happen next.

To ensure Serpa's victory, Samper relaunches, in the last eighteen months of his term, what he calls *el Salto social*, the social leap initiative. The content is concerned, these measures are irreproachable. Their true goal, however, is not disclosed. It can't be acknowledged, after all, that their purpose is to ensure future votes for the interior minister. For example, the government announces a plan to create vouchers ensuring access to medical treatment for the most disadvantaged people, but this plan is also a way to register and keep on file the names of group of people who would potentially promise to vote "the right way" in exchange for the vouchers. In the same fashion, the government creates vouchers for families to use to obtain school supplies—again, clearly targeting people who are ready to vote for anybody in order to receive these benefits. Similarly, retired people are enlisted with the offer of food stamps.

As if this wasn't enough, this "social leap" is also going to allow the Samper–Serpa team to divert, behind the scenes, considerable sums of money to finance Serpa's presidential campaign. They're going to carry out a veritable holdup of the country's finances. Denouncing this pillage becomes my main focus throughout 1997.

Once again, I'm virtually the only one in the legislature who tries to explain to Colombians what our government's sudden concern with social problems really conceals. How can one make people who have nothing understand that the pittance they are being offered today will turn into a poisoned gift tomorrow? How can one make them see that these piddling programs will do nothing to relieve the poverty and exclusion from which they suffer? On the contrary, because this "social leap" is merely a pretext for diverting monumental sums that will have to be paid back as soon as the elections are over through more taxes on all of us. My room for maneuvering is very limited. In each of my speeches, I risk appearing to be against everything, and, in this particular case, to be rejecting a social windfall desperately needed in the poor neighborhoods.

In the fall of 1997, the debate on the extradition of drug traffickers returns to public attention. It has been suppressed by Samper and his people for a long time, but the United States is getting impatient, and the pressure is considerable. Samper can't officially declare himself against extradition; that would amount to a confirmation of international acceptance that his campaign was financed by the Cali cartel. Instead, he says that he's in favor of reestablishing extradition if the legislature votes for it. But he's thinking about his allies, the Rodriguez brothers, who are still in prison. There can be no question of handing these men over to the United States. Betraying the pact he's made with them would be tantamount to signing his own death warrant. How can he save the Rodriguez brothers and at the same time appear to play by the American rules. It's simple—he can declare that the extradition treaty will not be retroactive, so that it cannot be applied to those who are already in prison. This, again, is a clean way out.

The legislators, the vast majority of whom are also financed by the Rodriguez brothers, grasp with relief this miraculous lifeline Samper has thrown them when he proposes that the treaty be made nonretroactive. For them as well, sending the Rodriguez brothers to the United States would mean death. On the day of the debate, I am the only one arguing for retroactive, total, implacable extradition. As always when the interests of drug traffickers are at stake, I am facing a full house. These representatives, whose ties to the cartel are by now well known (I've published the receipts they signed), present, with their hands on their hearts and without blushing, resonant legal arguments against retroactivity, in the name of the battle against . . . Yankee imperialism! I point at them that it's easy to blame everything on Yankee imperialism, but beyond these oratorical jousts I clearly discern in all my colleagues, including those with the most integrity, a gnawing, terrible fear of the drug traffickers. Besides, there are only three of us who are going to vote for a broad version of extradition, thus bearing the burden of sending the Rodriguez brothers to prisons in the United States.

Samper will continue to pay his debts by submitting to the legislature an apparently anodyne bill. On the pretext that our prisons are overcrowded, he proposes a system that will allow prisoners to serve their terms "in their own homes." One might expect this clemency measure to be applied to petty thieves, but no, it's intended, strangely enough, for a certain category of white-collar offenders. On closer inspection, it turns out that the bill concerns various forms of illegal enrichment and, in particular, the crimes committed by those politicians accused in the trial 8000. I raise a storm. The people accused in the trial 8000 are for the most part representatives, and they would thus be set free by their own colleagues without even having been tried. Finally, I win. Thanks to pressure from the press, the project is rejected at the last moment by the Senate.


The year 1997 is the most painful of my life. Powerless, I witness the whitewashing of a corrupt state, the institution saving its own gravediggers, proving without a doubt that the whole state apparatus is infected, corrupt to the bone. What hope for a cure can we have when the man who's on his way up to succeed Samper is his clone, Horacio Serpa? For the first time, I wonder why I'm conducting this battle. And if I'm wondering, it's because I've now sacrificed everything to this battle. What good has it done? How has Colombia benefited from what I have achieved? I see it sinking deeper, inexorably, and, instead of saving it, I feel that I'm going down with it.

On some evenings I leave the legislative building completely discouraged. What am I going back to? An empty, horribly silent apartment. Most of the time, Juan Carlos is not even home yet. I go into the children's rooms and sit on their beds. I look at their things, plump up the pillows, and sometimes find the strength to smile at the memory of something they said or did. Sometimes my eyes fill with tears and I remain prostrated, wondering what I'm doing. Why have I sabotaged the marvelous happiness of our life together? To bring about change, to invigorate political life, to show the way? But I haven't changed anything; power is still in the hands of the same men. It's as if I hadn't said or written anything, as if I hadn't existed. I'm depressed, lost. How could I have done it? These men I'm fighting, who have cost me all my strength, are not worth one of Melanie's smiles, a single lock of Lorenzo's hair. And yet I have not hesitated, I've chosen these men over my children. Not for a moment, on leaving them in Auckland, did I question the meaning of my commitment. But now, six or eight months have passed, and there has been no decrease in the violence that's wracking Colombia, no letup in the threats made against me that would allow me to consider resuming my family life. Are my children going to grow up entirely without me? Overcome by pain, I wander around the apartment that I bought for them, decorated for them, finding, every door, reminders of them, of their absence.

Our telephone calls only intensify my awareness that they're at the other end of the world. One Sunday I call them, full of nostalgia for our Sundays in Bogotá, but for them it's already Monday. They have to get ready for school, they're late, they have only a minute to talk. It seems to me that we're always operating on different schedules, as if we weren't living on the same planet. And then, as the months go by, I no longer know what to say to them, and soon, it seems my words no longer reach them. They must fill the silence, true, but they're litanies that no longer awaken any spontaneous response at the other end of the line.

"Oh, Loli, honey, my darling, I'm so happy to hear your voice! How are you? What's happening at school?"

"I'm fine, Mama, but I have go."

"You don't miss me too much? Tell me, Loli, you're not unhappy, are you?"

"I've got a game. I was just leaving."

"What kind of game, darling?"

"A soccer game, Mama, of course!"

"Oh, yes. You're playing soccer, then."

"I've been playing soccer for a long time. Didn't Papa tell you?"

"Yes, yes, he may have, forgive me. And school, everything's going well, really?"

"Yes, Mama, you already asked me that a minute ago."

Finally one day Lorenzo gets fed up:

"Mama, why are you always asking me the same question: 'How is it going at school? How is it going at school?' School's fine. We always say the same things, it gets boring."

That day I understood that my calls annoy them, because we no longer had anything to say to each other, nothing to share. I didn't know anything about their friends, their interests. And they were still too young to understand what I'm trying to do. This time, I go away to cry.

But when the initial pain is over, Loli's annoyance actually comforts me. It's a sign that he's growing up, that he wants to live, to have fun, and that ultimately he's getting along fine without me. That's great!

Perhaps this is the most difficult and painful thing a mother can have to do: to accept the fact that her children are growing up happily despite her absence. I accept it; I have no choice. But sometimes it hurts so much that I have to cling madly to any event that can bond us, despite the six thousand miles separating us. When I find out Melanie is learning to play the piano, and that according to her teacher she's exceptionally talented, I decide that I absolutely ought to have a piano in my apartment. I immediately set out to make this financial disaster of a purchase. I consult specialists, I inform myself, excited to be finally making a contribution to my daughter's education.

Melanie is eleven years old. Very soon, she's going to become an adolescent, a little woman. It seems to me that she can't go through this without me. How can she go through this without me? It would be painful for her, and for me. This thought obsesses me to the point that in my most delirious dreams, it comes down to this: Colombia has to emerge from this nightmare so that Melanie can come back. It absolutely has to—and right away, right away!

Colombia is getting along no better, but the children and I discover the Internet. That puts an end to those blind, distant, hopelessly futile telephone conversations. Now we speak to each other on the screen, we smile at each other, and a whole forgotten intimacy is reborn. Our almost daily on-line rendezvous soon determine our schedules. For me, they take place between two and three in the afternoon, and from this point on, I'm always arriving late at the legislative sessions. The immediacy of the contact stimulates the children's interest. I see Loli in his soccer outfit, Melanie in her room, photos of their friends. I hear about the books they're reading. For Melanie's twelfth birthday, I throw a party. My parents come; we've filled the room with balloons and bought a cake. When she appears on the screen, we see her break into laughter, delighted and abashed. My mother reaches out her arms to Melanie over a dozen candles.

"I'm going to blow them out for you, darling, and you'll give me your piece of the cake, okay?"

Melanie and Lorenzo bring the cake Fabrice got for them and we celebrate together. Really together. At this moment, bound by the same wild laughter, we forget that an ocean separates us.

My children get to know their cousins, the children of my elder sister, Astrid, over the Internet. New bonds are formed. Anastasia was born shortly before they left for Auckland, now they want to see her grow up. Sometimes we spend two hours laughing with her. When Stanislas is born, he too is introduced to my children over the net, and I can see Melanie's emotion, Loli's astonishment.

This artificial proximity sometimes leads to difficult moments. My children are there, as if I could reach out and touch them, but if they are confronting any type of danger, there's nothing I can do for them. One day I connect with them, all the more impatiently because I know Fabrice has gone to France. Melanie and Lorenzo have been left with Lise, their nursemaid from the Seychelles, in whom I have complete confidence. But Fabrice's absence worries me all the same. Melanie answers, apparently happy to talk to me, and yet I sense that she's troubled.

"Are you all alone? Where's Lorenzo?"

"He's busy."

"What do you mean, busy? Call him, Melanie, I want to see him at least for a moment!"

"That's just it, Mama, he doesn't want you to see him."

"That's crazy! What's going on? Loli, come here to the camera, my darling . . ."

"He doesn't want to, Mama."

"Melanie, please, go get your brother. What's happening, anyway?"

Lorenzo appears, and I seem to be sinking into a nightmare: he's all puffed up, unrecognizable, his face covered with red bumps, his body grotesquely swollen.

"Loli! What's happened to you! What did you do?"

"It's an allergy, Mama. Lise was taking me to the doctor when you called."

"What doctor? This is serious, Loli. Let me talk to Lise, quickly."

A pause, then Lise comes up to the screen.

"Listen to me, Lise. Get a taxi and go immediately to the hospital. Don't go to a regular doctor, he won't be able to do anything. Go to the emergency room. Leave right away and call me when you get back, I want to know everything that happens."

Then, from Bogotá, I call half the world—Fabrice's secretary in Auckland, his office in Paris, three allergy specialists in Bogotá, then the hospital in Auckland, then Paris again. Loli probably has a kidney problem, or so I gather from the doctors' veiled words. My heart is beating so hard it feels like it's going to burst. I panic, crazy with fear and helplessness.

On this day, I don't leave my telephone. I can't do anything until I get reassuring news from the hospital: Loli is being treated, he's out of danger, the doctors are confident that there's no longer any reason for concern.

Finally, during the summer of 1998, all of us meet in the United States. Day and night, I've been looking forward to this first family vacation after eighteen months of separation. I've prepared for it with excitement, convinced, like all guilty parents, that a large number of gifts would magically make up for all the love I've not been able to give. I've gone to all of Bogotá's most fashionable shops to buy clothes for my children, to spoil them, to see their faces light up at the sight of so many wonderful things. So I arrive, with two heavy bags, and my heart overflowing. Loli is ten years old, still a child; everything looks good on him, and he likes everything. But very quickly I see that as far as Melanie is concerned, I've done something wrong. My daughter watches me unpacking the presents with an expression of growing displeasure, a certain annoyance. In my loving blindness to accept it, I make things worse by saying:

"Look at this, my sweet. I'm sure you're going to love this."

Melanie looks, and then suddenly says:

"Mama, I stopped wearing T-shirts with cute animals on them a long time ago! You haven't realized that . . . that . . ."

"She's right. Melanie is bigger than we thought, Ingrid," Juan Carlos says softly.

And suddenly I see Melanie as she is, no longer the child I strove to hang on to with all my strength. And I'm embarrassed, unhappy— so unhappy that she's growing up without me, that she has also left me by the wayside.

# CHAPTER TWENTY

IN THE MEANTIME, Colombia has resumed a place in my life. I no longer have any doubts about the usefulness of the battle I'm waging. In the meantime, I've been elected to the Senate, I've created my own political party, and I tell myself that if I can someday, in some way, make this country into a real, open, generous democracy, I would have contributed to my children's happiness in my own way.

The year 1998 is approaching. The legislative elections will be held in March, and my ambition is to be elected to the Senate. In Colombia, a senator has more authority than a representative, and I need that extra authority to expand my audience, to continue my battle. But I no longer want to run under the Liberal Party banner, the one flown by Samper and Serpa, the men who have brought the country to its knees. The Liberal Party, whose code of ethics I drew up four years earlier, has not dared to exclude me, despite the blows I've struck against it. To expel me would be to admit that ethics plays no role in the party. I've made up my mind to leave the party, but I don't want to join its old adversary, the Conservative Party, whose representatives are hardly any better. It's time to create something else, another political family for people who share my beliefs. If I'm elected to the Senate, it will be for them, to speak on their behalf.

It's Juan Carlos who, one evening, first suggests the idea of creating a separate party. "You're talking about creating a political party, Ingrid," he says to me. "And that's what you have to do. At this point, you can't continue unless you create your own group. Look, you're not alone . . ."

"You need fifty thousand signatures to start a political party," I reply. "That's impossible, especially in only two weeks!"

"I'm not so sure of that. Get your supporters together, and think about it. Right away. If you decide to go for it, I'm with you. I'm enthusiastic about the idea of a new party, and I think that after four years of Samper, it's now or never."

He's right—it's now or never. The group of twenty faithful supporters who gather at my home that same evening also agree. Thanks to their excitement, the idea of gathering fifty thousand signatures throughout the whole country in only two weeks no longer seems implausible. We have devoted followers in every town; it's up to us to mobilize them, to get the administration's petition forms to them as soon as possible. Every person of voting age has the right to sign the forms. We have to count on the fighting spirit of our volunteers and on my popularity, for only people who have followed my career will get involved.

And it works! A month later—just before the deadline—we have nearly seventy thousand signatures. The door has opened, and we've created a new political party.

The news delights Juan Carlos, who immediately goes much further than I had ever expected: he reorganizes his public relations office, the campaigns he's running, his clients, and, in an extravagant, magnificent gesture, devotes himself to my cause.

But other than the signatures and the hostility of the political class that looks at us with anguish, we have none of the elements necessary to construct a political party. We don't have the organization, the activists, the offices, or, especially, the money. Juan Carlos doesn't see this as too much of an obstacle; he has confidence in me, and in everyone who has supported me for the past four years, and in those who will join us as soon as we raise our banner. And speaking of a banner, the first thing to do is to give our party a color, a name, an identity that will make people recognize us and want to join us, to build something together, to dream.

"Juan Carlos, find me a word that immediately says who we are and what we want. I want something different. I don't want anything that reminds anyone of what already exists: the Worker's Party, or anything of that kind, it's sad, lugubrious."

One day, as we are driving back from the market, Juan Carlos suddenly slaps his hand on the steering wheel.

"Ingrid, I've got it! I've got it, the name for your party: 'Oxygen'!"

He stops at the side of the road, happy, ecstatic. The word immediately grows on me.

"That's it! That's exactly right, Juan Carlos! Oxygen! It says it all. First of all, it speaks for the environment, but also it points out the suffocation imposed by the other parties, and the hope for release that we bring. The hope for life. It's a magical, exquisite word."

Ablaze with excitement, he says:

"Oxygen, that's you, Ingrid. I can already see the posters. We need an ultrasimple message: "Ingrid is oxygen." Period. We're wallowing in shit in Colombia, we can no longer breathe, we no longer have any hope, any dreams, anything. This Samper guy has screwed us over completely. Our poster has to be perfect. It'll have to show you against a background of blue sky, the very image of hope, of openness, of youth."

That night, Juan Carlos doesn't go to bed. The next morning he brings me the first mockup of our poster, my photograph against an azure background, under the slogan *Ingrid es oxigeno*. He'll make *oxigeno* our party's motto, transforming the *x* into a small, cheerful, yellow man juggling a red ball (the dot over the *i*).

Now that the party has a face, we can begin mobilizing our activists, and especially our candidates. We will present candidates for both the House and the Senate in all major cities. Juan Carlos is determined to see to it that I'm not alone in the legislature to do battle with these bandits. In the next legislature, we can act as shock troops.

Soon the poster with the blue background appears all over Colombia. All of them have the same motto, *Ingrid es oxigeno*, but depending on the region, the poster also displays a picture of one or two local leading candidates. Juan Carlos creates a look for us, too: we all wear the same short-sleeved polo shirt, but in different colors, depending on the color of our eyes or our skin. Our campaign stands out in clear contrast to that of other parties.

In Colombia, candidates pay little attention to their official campaigns, simply because the important "work" is happening backstage. For them, the main point is to buy enough votes, neighborhood by neighborhood, using sinecures and promises. What do their posters matter? They always show the stiff candidates in ceremonial dress, in front of a bookcase. But our poster makes them look as though they were half a century behind us! The contrast is so great that people laugh; our casualness makes our rivals look even more rigid and outdated.

Along the same lines as the condoms against corruption that I handed out on the streets four years earlier, Juan Carlos comes up with two inexpensive objects that will embody our message: an anti-pollution mask and a bubble of oxygen. Our candidates will flood their regions with these two symbols: one of them says implicitly how bad the stench is, the other how strong our desire to move toward a cleaner world.

Now we have to design our election T-shirt, an indispensable tool in Colombian campaigns. The parties traditionally give them out to the people, who, with the little they have to wear, are only too happy to receive them. Juan Carlos creates a T-shirt consistent with our party, clean, modern, and stylish (it came out so beautiful that I'm still wearing it today). At first, it seems that we won't have enough money to print more than two thousand, a ridiculously small number when our rivals are handing them out by the truckload. We call a meeting of our headquarters staff. We have to make a decision quickly.

"Obviously, if we buy contraband Chinese T-shirts we can make ten times as many for the same price," one person says discreetly.

"I hope that's just a joke!" I say. "I remind you that the core of our platform is to battle corruption. Let's talk with people in the textile industry and try to get a good price. Otherwise we'll have two thousand T-shirts, and that's it."

I think we didn't produce more than a thousand, but oddly enough, our strict adherence to the law probably won us more votes than if we had printed up a hundred thousand. For something strange happened: a month before the election, leaders in the Colombian textile industry, which was being strangled by Samper's policies, took interest in the steps candidates proposed taking to help this sector, which is one of the pillars of Colombian industry. The candidates, of course, made all sorts of promises to the textile leaders. But cleverly enough, the industry leaders looked at the candidates' record and the inconsistency between their speeches and their actions. And what did they discover? That with the exception of the candidates associated with Oxygen, they all purchased contraband T-shirts for their campaigns from Asia.

Ten days before the vote, the scandal was on the front page of all the newspapers, embarrassing the traditional parties and helping us enormously. During that time, I recall one of those commentaries I heard on the radio that warm the cockles of your heart. "We can be confident that Ingrid Betancourt will keep her campaign promises. She's putting them into practice even before she's elected: not one of her candidates has resorted to contraband. Oxygen's T-shirts are one hundred percent Colombian!"

During the final weeks I'm constantly on airplanes, going to every town to support our candidates. It's exciting and exhausting all at once. For the first time, I'm getting a taste of what it means to campaign on a national scale. I go to Barranquilla, Cali, Medellín, Popayan, Cúcuta, and everywhere people turn up, in high spirits, wearing our antipollution masks over their noses. They kiss me, they take snapshots of us surrounded by balloons. I speak before friendly crowds who believe in me, in us, in our party. With each new meeting, I'm making headway. People are regaining confidence, I'm sure of it. But how many supporters do we really have? Enough to open the doors of the Senate? I no longer know. Some evenings, I huddle alone in my hotel room, drained, exhausted, and even sick from moving back and forth on a single day between the tropical temperature on the coast during the day and glacial mountain temperature at night. I have strong doubts. How important, I think to myself, will these crowds be in comparison to the neighborhoods and suburbs where our politicians are buying votes with millions of pesos without even having to move? Is it ridiculous to play the democracy game all by myself? Am I being naïve and stupid too?

I can still hear the president of the House, who's also seeking a seat in the Senate, telling me in the corridor: "Representative from Bogotá, okay, Ingrid, you got lucky. But you're dreaming, old girl, if you think you're going to get into the Senate. Only big people, seasoned politicians get into the Senate. Just look at my region, Santander, which I know like the back of my hand. How many votes do you think you'll get there? Offhand, I'd say three thousand, at the most. And that's a generous estimate. It'll be like that all over the country. Don't kid yourself. If you get twenty-five thousand votes in all, it'll be a success, but it won't be enough to get you elected. It's impossible! Impossible!"

For years, he's been preparing his run for the Senate. And I know how, because he makes no bones about it: by giving local barons trips to the four corners of the earth at the expense of the legislature, in exchange for five hundred votes per trip. It's up to the barons to find the little gifts that will persuade voters to cast their votes for the feudal lord. And I'm not offering anything, not even a T-shirt, not even a sandwich, only promises, words of hope, words, always just words. If I had nothing to feed my children, would I go to applaud Ingrid Betancourt instead of voting for X or Y in exchange for a little job or a free meal for the whole family? Some evenings I don't have confidence in anyone, not even myself.

The great day arrives, and as usual it rains, a cold drizzle. As soon as the polls open, Juan Carlos and I leave to make the rounds of the Bogotá polling places. We're accompanied by a photographer from *El Tiempo*. It's Sunday, and the streets are empty, dreary, and gray. I'm very worried, and our first visits do nothing to relieve my concern. People don't greet me, not even a smile; they don't even seem to recognize me. "Shit," I say to myself, "if that's the way it's going to be all over the country, we're done." I glance at the photographer; he must think he's wasting his time. A terrible morning. Juan Carlos also seems worried, and, as usual, he doesn't try to make things look better than they are. Are we going to fail just when Colombia, drained and discredited on the international scene, so badly needs to recover its footing? And who except people like us can give it that last chance?

At four in the afternoon, the polls close. A few minutes later, the radio makes projections for Bogotá on the basis of the first vote counts. Our worst fears are confirmed: I'm not even named among the candidates in the race. Juan Carlos and I look at each other, discouraged, incapable of saying anything to each other. Silently, he drives towards the Registraduria, the building where the results for the whole country are consolidated. At least there we'll have, hour by hour, the numbers that reveal the extent of my defeat.

On the radio, the news bulletins continue to be depressing. We go into the building discreetly; dozens of journalists jam into the immense hall where the terminals have been set up. City by city, polling place by polling place, the screens report the results. We stand in front of the elevators, a few yards away from the press. I want to retreat to a calmer room, on the next floor. Suddenly, a reporter sees me and a group begins to move toward me. Photographers and cameramen bump into us, throw their lights on us, and hold out microphones in my direction.

"What do you think of the first results?"

"Nothing, I don't know what they are. I've just listened to the radio, and apparently . . ."

"*Doctora*, you're in the lead, one of the three top vote-getters."

"What do you mean, one of the three top vote-getters?"

"Right now, you have one of the highest scores on the national level. What do you have to say?"

Juan Carlos and I exchange a stupefied look.

"Wait," I say, "I haven't seen the results, I just got here. Where are the numbers? Give me time to see what's going on."

The reporters step aside, and others push me along as the cameras follow the screen that shows the list of candidates who have, for the moment, received the most votes. And then I stop dead. I'm not in second or third place, but in first! There's no doubt about it, that's my name at the top of the screen. I'm overcome by emotion. I feel like kissing Juan Carlos, and I want everyone who's been fighting at my side day and night over the past weeks to join us. I feel like crying: we've won, and Colombia has won! Silent for four years, it's now expressing a splendid disavowal of Ernesto Samper and the legislature that absolved him. It's showing that it believes in us, in me, that it has confidence. I try to catch my breath, to remember that a merciless political battle is taking place before our eyes, and that from then on, I will be the head of a party.

"Listen," I say, my voice a little unsteady, "it's only five-thirty, everything can still change, let's wait a while and see how the numbers go."

Something tells me that these men who've tried to assassinate me won't let me win so easily. Suddenly this first place seems too great a loss of face, too insulting for them and for the system they embody. A terrible fear eclipses my first moments of happiness. They control everything, they control most of the people who are counting the votes, and they're going to try to steal this victory from us, I'm sure of it.

I promise the reporters I'll be there when necessary, and I go to join Juan Carlos. My intention is to follow the returns city by city. All I have to do is click the button and the computer gives the cumulative vote for each candidate in real time. We sit down in front of a terminal.

It's six in the evening, and almost half an hour goes by without a single problem. Then, the returns from Cali suddenly stop coming in. While everywhere else the figures keep rising, the ones from Cali don't budge.

"Juan Carlos," I say, "let's go up to see the *Registiador.* This isn't normal, I'm scared stiff. As if by accident, Cali just happens to be a problem."

The Registrar is in his office, surrounded by about twenty people.

"What's happening? Cali is no longer transmitting results."

"Really? Wait . . . yes, you're right."

"I want to know why."

"I'll call them right away, *Doctora.*"

I listen in on the conversation. He nods and hangs up.

"They've had a power outage, no reason for concern."

"What do you mean, a power outage?"

"They say wires are down, *Doctora.* There's been a strong wind in the region."

I have my cell phone. Without taking the trouble to respond to him, I call our people on the scene.

"Eduardo, it's Ingrid. What's going on?"

"They've closed the Registraduria, and they're not letting anyone in."

"What's this story about a power outage?"

"There's no outage, the lights are working perfectly."

"You're not having a storm, then, no wind?"

"Not a breath of wind, Ingrid, why do you ask?"

I hang up, and this time I explode:

"Listen here, there's neither wind nor a power outage in Cali. It's obviously a ploy to conceal fraud. I'm warning you, I was leading in that region before the interruption, and if my votes decline after the returns start coming in again, I'm going to inform the reporters."

He calls back, makes a show of doing something. I hear him say I'm in his office. Is he part of the plot? I don't know, but obviously I have no confidence in this man.

When the results start coming in again twenty minutes later, the trend has completely reversed. I had about fifteen thousand votes in the Cali area when the reporting was interrupted, but for the rest of the night, I don't get a single additional vote. Of course the votes for the other candidates continually increase.

A month later, employees at the Registraduria tell me, in strict confidence, that with the help of certain officials, about forty thousand votes were stolen from me that night. If I hadn't gone up to the Registrar's office, I might not have won my seat in the Senate.

That doesn't happen. In fact, despite the fraud, when all the votes are counted, I receive more votes than any other candidate in the country. It's an immense victory.

At our campaign headquarters, I find a delirious crowd. The building is lit up and crowds are spilling onto the street. When I arrive, people shout and weep; many of them throw their arms around me, hug me. It's difficult to make my way through the crowd, and it takes me forever to get inside. I want to thank all these men and women who fought for us and came here when our victory was announced. As I finally come into the room, musicians—whom someone has brought there or who came spontaneously, I don't know which—

strike up the national anthem. My parents are there. Mama embraces me, weeping. Later, the musicians play Colombian folk music, and my father, deeply moved, offers me his arm.

"Dance with me, my dear, and we'll start the ball together."

The evening turns into a celebration for and with the people that lasts until the next morning. In the course of that night, I see many old friends from the French school with whom I'd lost contact. It's as if the people whom I love deeply and those I am fond of wanted to show me that this time they recognize me as someone worthy of having an influence on the country's fate.

Finally, I succeed in reaching Fabrice in New Zealand. He is moved.

"Let me talk to the children for a minute, I wish so much they were here in Bogotá."

"They're at school, Ingrid!"

"Excuse me, I forgot, I'm completely overwhelmed by events here."

"I'll tell them the news right away, in their classroom. They've been waiting for your call."

Eight years ago, Fabrice and I separated in Los Angeles. Today, the wound has healed. We have each found our own path, and we've not lost each other along the way. This is also a victory. Fabrice is the best father in the world, and he has once again become a precious companion.

The day after the election, I'm on the front page of all the newspapers. The success of Oxygen, and my own success in particular, is the real surprise of the election. Inevitably, I will become important for the presidential candidates. They are just now beginning their campaigns, and will try to obtain my support.

# CHAPTER TWENTY-ONE

**TWO CANDIDATES** seem to be positioned to win the presidency. One is Horacio Serpa, Ernesto Samper's faithful right-hand man, running as the Liberal Party candidate. The other is Andrés Pastrana, the Conservative Party's candidate. When he lost to Samper four years earlier, Andrés Pastrana fled Colombia after having made public the notorious cassette on which the Rodriguez brothers were heard praising Samper. At the time, Colombians wanted to believe in their new president's integrity, and they rejected Pastrana, a man who threatened to reveal new scandals. But now that Pastrana has returned to Colombia after a long exile, many people are saying that his only crime was to have been right too early. They recognize that they're indebted to him. But is it enough to beat Serpa? Apparently not. The system of allegiances, which Samper reinforced throughout his term, allows his designated successor to take the lead in the polls.

Because two candidates have an even chance for winning, Oxygen becomes an unavoidable force on the political chessboard. We're in a position to determine the outcome of the duel between the two candidates, and perhaps to ensure the victory of one over the other. We think this is an opportunity to obtain, in exchange for our support, strong commitments on issues important to us.

On the day before my election to the Senate, Andrés Pastrana calls me.

"Ingrid, let's have a quiet meeting. I'm sure we can work together."

"I don't know about that. We'd be risking our credibility by allying ourselves with a traditional party. If we do it, it would have to be in exchange for radical changes in the political life of the country, and I'm not sure you're prepared to work for those changes."

"Let's talk about it. I'm prepared for major upheavals. We agree about at least one thing: Colombia can't go on like this."

We make an appointment for a first, informal meeting at my home. Andrés Pastrana has a lot going for him. He's well aware that we'll never negotiate with Horacio Serpa. Besides, we're linked by many bonds. Andrés Pastrana is a very old friend of Juan Carlos, my husband, who campaigned for him in 1994. Andrés's brother, Juan Carlos Pastrana, is one of my best friends. We met in Paris, at the beginning of the 1980s, when I was at Sciences-Po. A brilliant journalist, he was then creating a foundation for democracy in Colombia, using German funds in particular. And the father of Andrés and Juan Carlos, Misael Pastrana, president of the republic from 1970 to 1974, was a close friend of my father's and was often a guest at our home when I was a child.

On the appointed evening, I come home late and find Andrés and Juan Carlos sitting in the living room, joking.

"I was trying to convince your husband to persuade you to join me," Pastrana says gaily to me.

"That's not true, Ingrid!" Juan Carlos interjects. "I don't intercede for anyone. Besides, I'm leaving now. I can't do anything for you, Andrés."

And Juan Carlos disappears.

"Ingrid, we have to stop Serpa," Pastrana says to me gravely. "You've acquired considerable influence on political opinion, and I need your help."

"I'm well aware of that, but many things separate us, in particular the infernal practice of buying votes. People in your party do it, and we'll never accept that. That's a precondition for any discussion."

"Ingrid, I'm trying to stop it. I'm the first to have suffered from that kind of rigging; I remind you that in 1994 I'd have beaten Samper if he hadn't bought half the votes with the Rodriguez brothers' money. If there's a victim of this system, it's me. Now it's true that some people in my own party are corrupt to the bone, but if I act too quickly they'll go over to Serpa and I'll no longer have any chance of winning. Trust me, my strongest commitment as president of the Republic is putting an end to clientelism.* In my government, people will be appointed because they are professionals, not because they have the back up of a congressman."

Andrés Pastrana goes away, obviously optimistic. He has given a convincing argument, shown himself to be very open, and played the simplicity card even in the way he came to see me: although he's in the middle of an election campaign, he came without a bodyguard, wearing casual clothes like an Oxygen candidate, spending a long time with me even though every moment is precious to him now. His message is clear.

Over the following days, an intense debate takes place within Oxygen's ranks. Let's suppose Pastrana is elected president; what would we want him to do in the first hundred days of his term to make a radical change? After much discussion and debate, we finally come up with ten points. Ten reforms are to be promulgated immediately in order to establish a genuine democracy in Colombia: electoral reforms, constitutional reforms to guarantee the independence of our institutions and especially of our judicial system. Since we know that the legislature is deeply infiltrated by corrupt elements, we demand

---

*Clientilism is a form of corruption, when government appointments are made because of a congressman's support. Once in office, the mission of these people is to rob in order to finance their boss's future campaign.

that this platform be adopted by referendum. We are certain that the people, unlike the legislators, will endorse it almost unanimously.

Eager to hear from me, Pastrana has even left me the number of his cell phone. As soon as we've decided on our demands, I call him.

"Ah, Ingrid! Well?"

"We have a proposal to make to you. When can you come by?"

"This evening."

"Fine, seven o'clock at my apartment."

I present our platform to him, and without a moment's hesitation he replies:

"I'm in complete agreement. I intended to make these reforms in any event."

"Wait, Andrés, I'm not asking you for vague agreement. We want these ten points to be adopted by referendum within the first three months of your term."

"I've understood that perfectly well, and I repeat that I'm in agreement. Moreover, just to make things completely clear, I propose that we sign a public pact to which Colombians, and my voters in particular, will be the witnesses."

A few days later, Andrés Pastrana sends me two of his closest collaborators, including his future minister of foreign affairs, Guillermo Fernández de Soto, to finalize the terms of our pact. The ten measures are to be presented to the legislature within the first thirty days of Pastrana's term of office. If the legislature rejects them, the head of state promises to submit them to the country by referendum within a hundred days.

On May 6, 1998, before an impressive crowd of journalists and television cameras, Andrés Pastrana and I sign the pact.

The next day, I throw myself body and soul into the battle to ensure Andrés Pastrana's victory. It's high time: we're only a month away from the first round of the presidential elections, and Horacio Serpa is leading in the polls.

Pastrana takes me along with him, and from then on we open all his major campaign appearances together. I discover the incredible madness of a presidential campaign, the fever, the battle with time, the endless flying back and forth, the midnight staff meetings. But I also discover the hope that my presence arouses in the Colombian people. Naturally, people give Pastrana an ovation, but it's different when they cheer for me; it turns into something that reminds me of concert halls, the delirious joy of the rising generation, tired of grandiloquent speeches, and receptive to sincerity, to hope, and my spontaneous stubbornness. Of course, I also derive a certain pride from this. But above all, I acquire the certainty that this Colombia is going to sweep away the one that has sold its soul to the drug traffickers.

But my presence begins to irritate the candidate's closest entourage. The crowd cheers for me, and its enthusiasm is arousing jealousy among the highest campaign officials. They see me as an intruder. They end up pushing me aside, and Pastrana makes his final appearances alone.

On May 30, Horacio Serpa comes ahead in the first round, but with only thirty thousand votes more than Pastrana (3,560,000 versus 3,530,000). The third candidate, Noemi Sanin, a woman faking the middle ground as an outsider from the two traditional parties. In fact, the changes she is proposing are an attempt to maintain the old corrupted system. Nevertheless she gets an amazing number of votes—2,800,000—thus opening the way for independent candidates in future presidential elections. Financed by big economic interests, Noemi Sanin is a sort of Trojan horse created by the ruling class in order to keep itself in power in a more presentable form. Sensing that the wind is turning and that the traditional parties are coming apart, Noemi is the seductive face that this corrupt system is now trying to sell to Colombians. She will have to guarantee the maintenance of the status quo and the privileges of those who are supporting her. Even though she has never been elected to any office, she will seek the presidency of the republic again in May 2002.

We have only two weeks to catch up. Pastrana and I decide to campaign separately so that we can cover the whole country. For me, this is exciting; I keep hammering on the terms of our campaign pact, and every night the crowds are there to hear me. But Pastrana is falling behind in the polls. And the more the polls show him losing, the more he waters down his speeches. Petrified, he loses all his audacity and leaves the catastrophic impression that his only goal is to be in agreement with as many people as possible.

One morning while I'm on the way to the airport, I can't stand it anymore, so I call him.

"Andrés, we're going to lose the elections, and I'm going to tell you why. You're no longer talking about the main point that will win us votes: the fight against corruption. You give the impression that you don't believe in anything anymore, even yourself. Do you know what people remember from your speeches? That you're afraid of losing, so afraid that your only program is to ask for help. But voters aren't going to help you; the last thing they want is a fainthearted president. We're coming out of four years of organized crime at the top, Andrés, and if you don't convince people that you've got the strength and the courage to set the country back on the right path, they're ultimately going to vote for Serpa. With him at least they know what they get, and he's doing a good job of playing on their antigringo xenophobia and chauvinism. That's all they have left."

Pastrana acquiesces. He's very worried, tense, but he assures me he's going to change. Will he find the inner strength to return to the battle? He has less than ten days to win back the undecided voters who will determine the outcome.

Ten days later, there's a transformation in him: a different man appears on television.

On the offensive, so aggressive that he risks upsetting people, Andrés Pastrana asserts that the first thing that has to be done is to wage war on corruption. "For nothing serious can be done in this country until we've done away with this system of corruption." Juan Carlos has had the ten points of our pact printed on the symbolic cover of a mock passport, and we've distributed this document throughout the country. Suddenly, Pastrana holds it up in front of the cameras.

"Here's our anticorruption passport," he cries. "This is the program that's going to help us win back for Colombia the place it deserves on the map of the nations. I promise to submit it to you by referendum."

The sudden change in tone of the campaign affects the mood of the country. People talk about crusades against corruption and call for disclosure. To those under forty, who identify with my proposals, Serpa's nationalistic posturing seems outdated, slightly grotesque, even absurd.

On June 21, 1998, Andrés Pastrana is elected president of the republic over Horacio Serpa by a margin of only 450,000 votes. A photograph immortalizing this victory shows the two of us with our arms around each other, waving to the crowd. Pastrana is wearing a suit, as befits a president, and I'm wearing jeans and a T-shirt, but our shared exuberance clearly shows the enthusiasm and hope that the country now feels at all levels of society. Moreover, whereas for months the polls have been showing Colombians' disenchantment, now, for the first time, 86 percent of Colombians say they're optimistic. Despite a catastrophic economic situation, hope has been reborn. People have confidence in Pastrana, and they say they believe he'll keep his promises. They're ready to follow him.

I'm well aware that I'm the guarantor of these promises. We all know that Pastrana's victory is largely related to our anti-corruption pact. I endorsed his campaign so it's up to me to see to it that he comes through on his commitments. Colombia will not put up with being deceived again. The risk is immense because, once elected, Andrés Pastrana finds himself at the confluence of two contrary currents: on one hand, the people's silent hope, and on the other the unscrupulousness of the traditional political class, which is determined to keep him from reforming anything at all. Oxygen foresaw this, and that's why we insisted that Pastrana resort to a referendum if the legislature, that seedbed of corruption, rejected the first set of measures.

Where would I be most effective? By joining the government, as Pastrana wants me to (obviously, he offers me a ministerial portfolio for Oxygen), or by speaking my mind freely without being part of the government? I'm convinced that accepting a ministerial portfolio for Oxygen would more or less force us to keep quiet out of fear of being considered disloyal, so I decide to refuse all offers.

The president must quickly name a commission entrusted with establishing the text of the reform bill that will be submitted to the legislature and, if need be, to the Colombian people. I eagerly accept a seat on this commission, alongside university professors and noted jurists. My attorney, Hugo Escobar Sierra, is also named to the commission, along with Humberto de la Calle, the man who supported me four years earlier when I was defending the Liberal Party's code of ethics.

The commission is already hard at work when Andrés Pastrana puts together his government. Curiously enough, he names as minister of the interior—a very important position in a country where there is no prime minister—Nestor Humberto Martinez, a long-time Samper supporter. As Ernesto Samper's minister of justice, Humberto Martinez was the man who so weakly opposed the *narcomico* before resigning in exchange for an appointment as ambassador to France. His sudden return to grace, this time to serve Pastrana at the highest level of government, seems a very bad omen to me. In my view,

Martinez is a man without convictions who's willing to swallow anything in order to advance his career. His nomination worries me all the more because the reform law we've almost finished drawing up will be submitted to him, and, as the minister of the interior, he'll have to defend it before the legislature. Is Pastrana laying a trap for us? Is this the beginning of the betrayal I've secretly feared from the start?

I call the president.

"I don't understand why you appointed that guy. The reform that's supposed to be the foundation of your presidency will be in his hands, and you know perfectly well that he doesn't have either the ideas or the courage."

"Ingrid, trust me. I needed a man who could win support among all factions in the legislature, and Humberto Martinez can do that. He knows the ropes and he's close to both the Liberals and the Conservatives. As far as the reforms are concerned, don't worry, I'll be shepherding them through personally."

I'm only partly reassured, and subsequent events do nothing to calm my fears. Very soon, in fact, Humberto Martinez undermines the reform commission of which he is a member de jure. Whereas recourse to referendum has the commission's support, the minister hints to journalists that a referendum is unlikely. This is obviously an overture to the legislators, and they eagerly accept it. They understand that Martinez is really there to sabotage reform, and that they have to support a man who'll help the old, corrupted practices to survive.

The commission itself does not see this. Focused on arriving at an agreement with the minister, it continues to work tirelessly. As a result, the prescribed deadlines are not met; September arrives, and the reform bill still has not been submitted to the legislature. Pastrana is well aware that if he delays too long, he'll lose his accumulated credibility. On September 8, he goes on television to emphasize his commitment to corruption and to repeat his promise to consult the people by referendum if the legislature rejects reform.

It's becoming clear that he's already lost the means of realizing his ambitions. The first signs of disbelief and disenchantment appear in political cartoons and in the caustic jokes that people bitterly exchange on every street corner.

There is, in fact, reason for doubt. In the days following Pastrana's appearance on television, the minister of the interior does his best to foment a rebellion against the referendum in the legislature. It's inadmissible, he says, to confront the legislators with an all-or-nothing choice (accepting the reform bill or rejecting it); they have to be able to change the text, rewrite it. This amounts to abrogating the pact I signed with Pastrana, and it will sabotage anticorruption reform before it even gets started.

Once again, I warn the president.

"Andrés, what's going on in the legislature is very serious. Nestor Humberto Martinez is manipulating everybody, he's deceiving us all. The commission is about to implode. If you don't take action soon, I'm afraid the reform bill is doomed, and along with it your commitment to the Colombian people."

He listens to me and proposes a meeting of a small committee with only three representatives from the reform commission: Humberto de la Calle for the Liberal Party, Hugo Escobar Sierra for the Conservatives, and myself for Oxygen. The meeting is set for September 19 at the Casa Medina hotel.

I'm expecting a crisis meeting in which we will rally around a president who has been weakened, but whose will is still intact and who is eager to regain control of the situation. We all know each other very well, and so we'll move to the central issue immediately— grimly, but confidently. At least that's what I imagine. But from the outset, the cards are stacked against me: Pastrana does not come alone, as he had suggested he would, but instead flanked by four ministers, his cabinet chief, and his personal secretary. I immediately sense a trap. Unfortunately, I'm right.

"I hope," Pastrana begins by saying, "that the reform envisaged is the result of a consensus among all the parties, and that it will be approved by the legislature. There will be no referendum."

Immediately I understand why he has had only the three of us come. Had he said this to the full commission, there would have been an outcry.

"Mr. President," I say, ceasing to address him in the familiar form, "you are about to renege on the agreement we made. It's never been a question of finding a consensus among the parties, which is impossible and could lead only to no change at all, but rather of an appeal to the people. Today you're telling us that you're not going to have a referendum."

At these words, I watch his face go red. Without letting me finish, he jumps to his feet, pounds his fist on the table, and gives me a furious look.

"I won't allow you to speak to me in that way! I never intended to conduct this reform against the traditional parties, and besides, I promised to pursue a policy of reform, but I never promised to subject the country to a referendum."

It's a terrible, unprecedented scene—the president of the republic shouting, beside himself, enraged, his face scarlet, the veins in his neck protruding. His entourage seems to be encouraging him to speak to me in this way. Instantly, I understand what is going on. This man is afraid of betraying me, otherwise he wouldn't have come with this group of supporters. He knows perfectly well that he's going back on his word, and that's why he's angry. Angry at himself, of course, but also at me, because I've been harassing him for the past month, through the newspapers. Not a day goes by without the press mentioning my determination to see to it that what the journalists now call simply "Ingrid's referendum" takes place. Just imagine what this kind of bludgeoning means to a man who's so full of himself! A week before this meeting, in still another article devoted to "Ingrid's reform," *Semana* has presented Pastrana as a naïve guy who lets himself be led by the nose. Nothing could be more effective in getting him to head in the other direction.

Andrés Pastrana is certainly vain, and he finds my frankness and informality difficult to bear, all the more so because since his teenage years he's suffered from a deep-seated inferiority complex. Andrés was a mediocre student, lacking his brilliant older brother's intellectual and cultural credits. He also became a journalist—with his father's help. Their family owned a television information service for which he served as a newscaster—and it was because he was a television star that he was able, without too much difficulty, to become Bogotá's mayor.

I'm thinking all this while Pastrana continues his extraordinary tirade, which only confirms, unfortunately, the lack of stature of the man who has been called upon to decide the destiny of Colombia until 2002. For the second time, I have the terrible feeling that I've been deceived, as if history was repeating itself. In 1994, I wanted to believe in Samper and in his claim to be interested in social problems. Then, despite my deep hesitations, I supported him over Pastrana, whose limits I knew well. This time, I supported Pastrana against Serpa, without enthusiasm, but on the basis of a pact signed before the Colombian people—the same people who claim that in our country we are always forced to choose between plague and cholera. How can now one say they're wrong? Powerless to keep his promises, Pastrana will break them right in front of us, on September 19, 1998, by defying me like a pitiful child, stamping his foot with rage.

"People say she's leading me by the nose," he seems to be saying to his ministers. "Well, now you see who's in charge!"

I'm convinced he needs this scene to salve the wounds to his pride.

But the worst is yet to come. When we separate, a moment later, Pastrana takes me aside, and hugging me, says:

"Don't worry, Ingrid. Everything's going to be all right. You'll see, we're going to go through with this reform."

That evening, I come home a wreck. I've hardly closed the door when I break into tears. I weep with rage, without being able to stop, as I've never wept before. I feel I've been betrayed, used, mistreated. Once again I think of the students I met in Medellín, during the campaign, who told me: "You're going to get taken, Ingrid. This guy needs you, but he's just like the others." But I defended Pastrana, on the pretext that he'd also been a victim of corruption. I wanted to believe that through him we'd find a way out.

This betrayal, one of the most painful in my political life, leads to the conviction that one day I will have to seek high government office myself if we are to save Colombia from the corruption that is killing it. I become convinced that no compromise is worth anything if it's made with representatives of the traditional political class.

The very next day, Andrés Pastrana makes his turnaround official by publicly announcing that he's going to meet with all the political leaders in order to take the first steps toward a "consensual reform." A referendum is no longer mentioned, and the reform commission is relegated to the rank of a group of experts whose task is to provide material for the elected officials to consider. And so we witness the return of the corrupt old guard. For anyone still inclined to doubt this, the name of the first man invited to meet with the president comes as a brutal revelation: Horacio Serpa! After having promised so much, after having given people hope, Pastrana is opening the door to the devil.

What does he hope to achieve by announcing that he has invited me to meet with him the same day? Is he trying to fool Colombians? Or to obtain my backing for a reform from which clearly nothing will come?

The meeting is set for September 25. Shocked by the way Pastrana humiliated me at the Casa Medina hotel, the whole staff of Oxygen decides to accompany me this time. Our arrival at the Nariño palace causes a sensation, all the more so because Horacio Serpa and Noemi Sanin are already in the president's office. Andrés Pastrana refuses to receive us, and we find ourselves relegated to the office of his secretary, Juan Hernández.

"Fine," I say, "say hello to him for us, we're leaving."

"No, no, wait, don't leave like that, I'll tell him."

"There's nothing to wait for. Pastrana signed a reform pact with me, but now he prefers to deal with his former enemy."

Hernández senses a scandal; dozens of reporters are crowding around the doors of the presidential palace.

Pastrana understands what that means. He asks us to wait a few minutes.

Finally he appears, smiling, friendly.

"Ingrid! Come in a minute, we have to talk."

We find ourselves alone together—the last of my private conversations with him.

"What you're doing is very sad, Andrés. You succeeded in giving Colombians hope again, and now you're betraying them, betraying yourself. You'll never get the legislature to approve the kind of reform we dreamed of. You've chosen the worst possible course: you're betraying the people who supported you in order to form an alliance with your own worst enemy."

"No, Ingrid. Try to understand me, I want to reconcile the country."

"That's not what Colombians asked you to do. You're not here to search for a new consensus with criminals; on the contrary, you were elected to burn all bridges connecting you with that completely corrupt political class. We don't live in a European-style democracy in which all elected officials are more or less honorable, we live in Colombia."

The break is now complete. However, seeing me out, Pastrana adds:

"Ingrid, I ask only one thing of you: when you leave here, don't say that you've broken with me."

"What do you want me to say? That we're still together? No, I'm going to say that you are going off in your own direction, but I'm continuing to call for a referendum."

"Don't be angry with me."

"I'm not angry with you for myself, Andrés. But I'm angry with you for the country, very angry. You don't see what a historic opportunity this was."

I leave him, knowing I'll never return. Sad, to be sure, but convinced that I have to make my own voice heard.

Outside, the reporters are waiting. They rush toward me.

"We've just broken the election agreement that bound us to the president. The reform we wanted has been sacrificed to the most crooked politicians in Congress. Serpa and Noemi are going in through one door, and we're going out the other."

Eighteen months later, in March 2000, Andrés Pastrana, mired in deep corruption scandals, surprises everyone by announcing that he's finally going to keep his promises. He will put to a referendum a great political reform calling for the dissolution of the legislature and rethinking the election process. At the time, his ratings in the polls are at their lowest ever. This announcement alone causes them to rise by twenty points.

Naturally, on behalf of Oxygen, I give him my support, and tell everyone that this move is an act of courage. But both the Liberals and the Conservatives go on the offensive, and two months later, incapable of staying the course in this storm, Andrés Pastrana again reneges on his commitments. We'll have no referendum, he says, and we won't dissolve the legislature. This amounts to an admission that corruption has won. The Colombian people understand this, and now only 15 percent of them say they still have confidence in this tragically erratic president.

For us, the battle for the referendum continues, but over the months it becomes increasingly clear that this vainly hoped-for reform will be one of the main issues in the next presidential election, in 2002. For Colombia, four years will have been lost.

# CHAPTER TWENTY-TWO

UNFORTUNATELY, this is not the first time that Andrés Pastrana has resorted to heartfelt announcements that come to nothing. Like most of his predecessors, he's also made use of the tragic issue of the guerrillas. In Colombia, our leaders talk about peace whenever they are in trouble.

As he'd promised during his campaign—and this is probably the only commitment he will honor—the president has opened negotiations with FARC,* the main guerrilla group in the country, which has more than fifteen thousand men under arms. To his credit, in doing so he's shown that it's possible to enter a dialogue with the guerrilla leaders, it seems to Colombians that they're not blind monsters but men with ideas and ideals. Moreover, by conducting negotiations openly, before the television cameras, the president has succeeded in bringing our conflict to the forum of international opinion.

But by now the fragility of the "Pastrana method" is already apparent. A former television journalist himself, the president always

---

*FARC is the acronym for Fuerzas Armadas Revolucionarias de Colombia, or Revolutionary Armed Forces of Colombia.

seems to attach more importance to "media coups" than to serious reflection on the issues.

We immediately saw this when, in the euphoria following his election, he made a historic "gesture." In the name of peace, he granted FARC nearly seventeen thousand square miles of national territory. And what commitments did he get from the guerrillas in exchange for that? None whatsoever. This abandonment of sovereignty was made in the vaguest possible way, at the risk of sending the country the message that the state was ready to weaken itself in order to get into the good graces of the warlords.

The leaders of the various guerrilla groups remain coolheaded. At least this is my impression after having talked at length with them. They're perfectly aware that Colombian leaders use negotiations for electoral purposes, and that their desire for peace is not based on any long-term vision. So the guerrillas are pretending to want peace, and they have everything to gain by doing so. At the same time, they're preparing for war or waging it. For example, during the spring of 2000, FARC forces tried to smuggle ten thousand weapons into the zone granted to the guerrillas by Pastrana for peace talks. This scandal, in which the existence of a Peruvian pipeline came to light, ultimately brought down President Alberto Fujimori of Peru.

It's as if the political leaders and the guerrillas are helping each other along in order to maintain a state of war that suits them but is destroying our country. The guerrilla leaders don't want to be told that the battle they're waging in the name of the people is, paradoxically, strengthening the political class that is the source of the people's misery and sustaining the system of corruption under which it flourishes. Nevertheless, this is what I have told them. They know what I think, and this allows me to maintain distant but frank relationships with them, without ambiguity.

For the time being, negotiations have no chance of success. They've been warped from the outset. Their goal isn't to arrive at a conclusion, but to win time for the parties involved. Each of the adversaries is convinced it will eventually ensure its ultimate victory by military means. Everyone lies. And everyone pretends to believe the lies of the other.

In fact, the common people are the only ones who really want a negotiated peace. It's the common people who are burying their own every day, with thirty thousand deaths a year. To lower this number, it's not enough to be seated at a negotiation table.

Colombia is in a state of war. It is an internal, complicated, and extremely violent war that has claimed countless Colombians and destroyed our environment and economy. The war has also had a serious impact on other countries. Its most tangible effect is the drug traffic that has wrought havoc on the streets of America. I want to see Colombia free of war, with its democracy reclaimed, its peace assured.

This is why we first need to decide what kind of peace we are going to seek. Do we want a fake peace imposed by the use of terror? This is what the paramilitaries are offering us. Do we want a peace that results from a defeat of democracy and the installation of a communist regime? This is what the guerrillas are fighting for. Do we want a peace agreement negotiated by a corrupt regime that uses a false promise of peace as a tool to maintain the status quo that allows the select few to share the privileges? This is what our establishment is trying to preserve.

None of these possibilities will free us from the drug traffic emporium and the violence that accompanies it.

The peace that we, the Colombian people, want is a different one altogether. It is a peace rooted in the rules of democracy. It can't be a peace negotiated by a corrupt, feeble, and vacillating government.

People are tempted to think that a certain degree of corruption can be tolerated as a price for minimal political stability in the region. This was the view taken for decades, because corrupted governments and dictatorships were battling communism.

Today, however, the problem is more complex and more vital. Today, we are all citizens of the world. We are compelled to play by the same rules, and we all share the consequences of political turmoil, even if it seems to be contained within a specific place. What happens abroad has immediate consequences on our daily lives. This is what globalization is all about. If we want to make it work, if we want to prosper, we need democracy.

Today, we cannot expect to fight drug trafficking while we turn a blind eye to the corrupted ways of our government. These are, after all, one and the same, and they work together to make sure that things don't change.

Today, we cannot expect to enforce civil and human rights laws while we ignore the evidence of electoral fraud. This is precisely the kind of corruption that nurtures violence.

Our war in Colombia is the best example of this situation. In this war the Colombian government has permitted the creation of paramilitary forces. Financed by powerful landlords and drug traffickers, and too frequently trained with the help of high-ranking army officials, these illegal troops are used to confront the guerrillas, and they have an agenda of their own. These illegal troops are doing what the law forbids our army to do: they carry out massacres, tortures, and persecutions. Our government condones their victories because they hold back any form of subversion that challenges its authority—even though this tolerance ends up strengthening the very mafia the government claims to be fighting.

As long as the AUC (United Colombian Self-Defense) paramilitary forces are a clandestine instrument of the establishment, the Colombian government will lack the legitimacy required to discuss peace with anyone. In the end, the Colombian government accepts illegal money to win a war that protects not the lives of civilians, but the properties of those financing the war.

This is why peace cannot be sought without addressing openly the close ties between drug traffickers, paramilitaries, and guerrillas. Any peace process has to begin with a strong commitment from all parties to fight corruption in its most sophisticated manifestation—drug trafficking.

It is only when we target drug trafficking that we will truly weaken the financial supply channeled to corrupted politicians and terrorists and thereby arrest the perpetuation of violence that has crippled Colombia.

Three conditions are necessary to bring peace to our people:

1. The denarcotization of Colombia—We need to weaken the drug traffic partnership with terrorism by making a commitment to fighting it the sine qua non for any peace talks.
2. The enforcement of human rights laws—We need to reestablish government authority by severing the government's clandestine ties with the paramilitaries.
3. Support from the international community—We need partners to confront the corrupt and very strong Colombian political force in power.

There is no doubt that we need international support to accomplish these goals. The Colombian people cannot be left to confront such powerful organizations on their own. Their tentacles reach far beyond the Colombian people's control.

We will win this war if we can receive support from all over the world—and especially from the American people, not only because our two nations are the victims of this illegal drug traffic and the terrorism it finances, but because we need help from truly democratic countries in rescuing our own democracy.

There is a clear link between having a strong, legitimate democracy in Colombia and stopping the flood of drugs onto American streets. The American people need to know that.

If the drug lords are financing elections and the electoral fraud, they gain control over the government, parliament, and judiciary. But if we have clean elections, we will have a political turnaround. Without their accomplices in the government, the drug lords are like fish out of water. It will be easier for us to beat them, and at the same time it will shut the valve that fuels violence and war.

I believe strongly that only true democracy will give us, the people of Colombia, the means of defeating the forces behind the war that is being waged against us internally.

A strong legitimate democracy will induce the emergence of a new Colombia. A war against the economic and political power of drugs will cut off the paramilitaries and the guerrillas from their financial source. At the same time, a true, democratically elected government guarantees a political structure that will address and ensure people's desire for social justice. A combination of these factors can pave the way to peace negotiations and a truce with the guerrillas and the paramilitaries.

There is no quick fix, but I am convinced that reclaiming our democracy is the very first step toward peace, and the sole condition for a true alliance against drugs and against terrorism between the people of all nations.

# EPILOGUE

**WHEN I RETURNED** to Colombia in the early 1980s, Luis Carlos Galán, the man who embodied hope for Colombia, had just been assassinated. The country, exhausted by decades of violence and corruption, was once again being put to the fire and the sword, terrorized by drug traffickers' daily attacks—the notorious "bombing war." As a candidate for the presidency, Galán insisted that our salvation lay in ethics. Corruption, he repeatedly said, was the origin of Colombians' terrible suffering. I knew he was right, but I was not yet thirty years old and had no experience with power. Today, I've taken up the battle in the name of all those who have died without seeing the first rays of the dawn. For the dawn is there, really there, for us Colombians.

We've covered half the road. When I hear Papa telling me, "You know, I'm no longer Minister Betancourt, I'm Ingrid's father!" I hear his pride. It echoes that of a nation that wants to believe in me, that is slowly recovering its confidence after a century of lies and betrayals.

I will not betray that confidence.

Colombia has always been headed by little factional leaders. Our real leaders have all been assassinated. Mediocre men get themselves elected in order to enrich themselves, and then go off somewhere else to enjoy life. Where does Ernesto Samper live today? In a fashionable suburb of Madrid. "Leaders" like him never believed in my country, and they have a profound scorn for our people.

I am the opposite of them. I love Colombia, to the point of making the most painful choices possible in order to have the right to live here. I love our people because I know that having been the victim of the cruelest violence for more than a hundred years, we carry inside ourselves treasures of courage and passion. The anarchy of today's Colombia is a call for help that the world is refusing to hear.

This violence is the cry of all those who can no longer bear this bandit government, this outlaw government. It's also our shame. The guerrillas, the paramilitaries, the drug traffickers, the gangs of delinquents who are tormenting our country are even more barbarous than the disgraceful government they claim to challenge.

In spite of that, the great majority of us have refused to sign a pact with the devil. Doomed to live in a daily hell, we have not lost hope. We Colombians dream of peace, of harmony, of justice, and we teach our children to praise candor so that we don't lose what little bit of paradise is left.

With such treasures, it will not be difficult to construct the Colombia I dream of, that many of us dream of. In ten years I've learned a great deal, and today I feel strong enough to lead this process to a successful conclusion. Imagine what kind of country we would have if we invested in work, in production, in creation, in pleasure, in our families, the extraordinary energy we devote to death. We live in isolation, we're vulnerable, we mistrust one another. Our social fabric is seriously damaged. The only organized systems—and they're remarkably effective—are those of corruption. We have to turn things around; black has to become white.

I want to accomplish this.

If what I've been doing for the past ten years hadn't produced a response, I wouldn't feel qualified to make such a commitment. But

I have twice been elected with a remarkable number of votes, and today I feel that I'm ready to put a stop to corruption. I also note that the same politicians who hate me also ask me to endorse their proposals, because they know I'm credible, that, unlike them, I can't be bought. In a sense I'm forcing them to think that they could also be different. I'm forcing them to imagine the Colombia of tomorrow, the one we all deserve.

Now that I've arrived at this point, will they kill me, too? My relationship with death is like that of a tightrope walker: we're both doing something dangerous, and we've calculated the risks, but our love of perfection invariably overcomes our fear. I'm passionately in love with life, and I have no desire to die. Everything I'm building in Colombia, I'm also building so that I can happily grow old there, so that I can have the right to live there without fearing disaster for everyone I love.